T0191793

Nonlinear Physical Science

Laurent Raymond⑩, Centre de Physique Théorique, Aix-Marseille University, Marseille, France

Victor I. Shrira, School of Computing and Maths, Keele University, Keele, Staffordshire, UK

C. Steve Suh⑩, Department of Mechanical Engineering, Texas A&M University, College Station, TX, USA

Jian-Qiao Sun, School of Engineering, University of California, Merced, Merced, CA, USA

J. A. Tenreiro Machado⑩, ISEP-Institute of Engineering, Polytechnic of Porto, Porto, Portugal

Simon Villain-Guillot⑩, Laboratoire Ondes et Matière d'Aquitaine, Université de Bordeaux, Talence, France

Michael Zaks⑩, Institute of Physics, Humboldt University of Berlin, Berlin, Germany

Nonlinear Physical Science focuses on recent advances of fundamental theories and principles, analytical and symbolic approaches, as well as computational techniques in nonlinear physical science and nonlinear mathematics with engineering applications.

Topics of interest in *Nonlinear Physical Science* include but are not limited to:

- New findings and discoveries in nonlinear physics and mathematics
- Nonlinearity, complexity and mathematical structures in nonlinear physics
- Nonlinear phenomena and observations in nature and engineering
- Computational methods and theories in complex systems
- Lie group analysis, new theories and principles in mathematical modeling
- Stability, bifurcation, chaos and fractals in physical science and engineering
- Discontinuity, synchronization and natural complexity in physical sciences
- Nonlinear chemical and biological physics

This book series is indexed by the SCOPUS database.

To submit a proposal or request further information, please contact Dr. Mengchu Huang(Email: mengchu.huang@springer.com).

More information about this series at http://www.springer.com/series/8389

Dimitri Volchenkov

Editor

Nonlinear Dynamics, Chaos, and Complexity

In Memory of Professor Valentin Afraimovich

Editor
Dimitri Volchenkov ⓘ
Department of Mathematics and Statistics
Texas Tech University
Lubbock, TX, USA

ISSN 1867-8440 ISSN 1867-8459 (electronic)
Nonlinear Physical Science
ISBN 978-981-15-9036-8 ISBN 978-981-15-9034-4 (eBook)
https://doi.org/10.1007/978-981-15-9034-4

Jointly published with Higher Education Press
The print edition is not for sale in China (Mainland). Customers from China (Mainland) please order the
print book from: Higher Education Press.

This Springer imprint is published by the registered company Springer Nature Singapore Pte Ltd.
The registered company address is: 152 Beach Road, #21-01/04 Gateway East, Singapore 189721,
Singapore

Preface

The present volume on *Nonlinear Dynamics, Chaos, and Complexity* explores recent developments in experimental research representing a new step into our understanding of complex dynamical systems and the relation between their structure and functions. The volume is dedicated to the memory of our colleague Valentin Afraimovich (1945–2018), a visionary scientist, respected colleague, generous mentor, and loyal friend. Professor Afraimovich was a Soviet, Russian, and Mexican mathematician known for his works in dynamical systems theory, qualitative theory of ordinary differential equations, bifurcation theory, concept of attractor, strange attractors, space-time chaos, mathematical models of nonequilibrium media and biological systems, traveling waves in lattices, complexity of orbits, and dimension-like characteristics in dynamical systems.

The volume opens with a memoir of Ms. Alexandra Afraimovich, the daughter of Prof. Afraimovich. The rest of the volume represents an important collection of works presents recent advances in discontinuous and nonlinear dynamical systems, chaos, and complexity science, important to understand resonance interactions in nonlinear and discontinuous dynamical systems. The authors discuss analytical solutions, the effect of time-delays, decisions under uncertainty, human-inspired network dynamics, and urban spatial networks. Nonlinear Dynamics and Complexity invites readers to appreciate this increasingly main-stream approach to understanding complex phenomena in nonlinear systems as they are examined in a broad array of disciplines. The book facilitates a better understanding of the mechanisms and phenomena in nonlinear dynamics and develops the corresponding mathematical theory to apply nonlinear design to practical engineering.

Valentin Afraimovich was a generous, gregarious, energetic presence at the very heart of nonlinear dynamics and complexity science communities, all of which were transformed by his presence. We hope that the scientific community will benefit from this edited volume.

Lubbock, TX, USA

Dimitri Volchenkov

Contents

Professor Valentin Afraimovich

Alexandra Afraimovich

Valentin Afraimovich (April 2, 1945–February 21, 2018). A world-renowned scholar, doctor of physical and mathematical sciences. Valentine's work has brought him international recognition in the late 70 s, when he was still a young scientist from the city of Gorky, nowadays Nizhny Novgorod, Russia. Since 1991, Mr. Afraimovich lived and worked abroad, occasionally paying visits to his homeland in Russia.

Valentine Afraimovich was born in the small town of Bogorodsk, near the city of Nizhny Novgorod. His father Alexander Afraimovich was a well-respected agriculturist. In 1941, while still in college, Alexander enlisted as a volunteer to fight Nazis in WWII. In the course of terrible fights near Smolensk he was wounded in both legs and sent back home to finish his studies. Valentine's mother, Anna Kubyshkina, descended from a large upper-class merchant family that lived in Bogorodsk for many generations. She was drafted in the fall of 1941, right after she graduated

A. Afraimovich (✉)
Ensenada, Mexico
e-mail: sashavalentine21@gmail.com

© Higher Education Press 2021
D. Volchenkov (ed.), *Nonlinear Dynamics, Chaos, and Complexity*,
Nonlinear Physical Science, https://doi.org/10.1007/978-981-15-9034-4_1

from her accelerated senior year at the Moscow Medical Institute. Anna worked as a military surgeon on the front line, received several medals for bravery and in 1944 was allowed to return to her hometown of Bogorodsk where she began working as a doctor in a local hospital.

After graduating from high school in 1963, Valentine Afraimovich applied to the top school in the USSR—Moscow State University. He aced the entrance exams, but alas, the state-sanctioned anti-semitism was rampant in the Soviet Union, and Valentine, just like the majority of the applicants with Jewish last names, has not been accepted. The official excuse for his rejection was a substandard performance in physics—examiners told the future world-renowned mathematician that his approach to physics is too math-oriented.

Fortunately, during that time Gorky State University had a wonderful practice of waiving entrance exams for the best applicants who were rejected by the top Moscow and Kiev universities. Thus, inadvertently, Gorky became a haven for the talented youth from all over the Soviet Union. This gave the historically merchant city, nicknamed "the purse of Russia", a reputation of being a relatively liberal and sophisticated provincial hub.

Valentine graduated from the Gorky State University in 1968. During the course of his studies he began research work under the direction of Leonid Shilnikov. Upon graduation he received a position at the research institute of applied mathematics and cybernetics, in the department of differential equations, where he has actively continued the work begun with Shilnikov.

An important stage in V. Afraimovich's work was his decisive participation in a classic work under the direction of Leonid Shilnikov on the structure of strange attractors in Lorenz system. Edward Lorenz asserted that long-term weather forecasting is impossible because the climate possesses the properties of dynamic chaos. In such systems the slightest deviation from the initial conditions leads to significant changes—Lorenz originated the term "the Butterfly Effect", which is erroneously credited to Ray Bradbury for describing the effect in his short story "A Sound of Thunder". Lorenz's findings were long shunned by the scientific world, but Valentine Afraimovich and his colleagues produced mathematical proof of Lorenz's theories. Their work was first published in a Soviet scientific journal and later translated into many languages. This publication brought international recognition to the group of young scientists from Gorky.

In two years, another collaboration by Valentine Afraimovich, Mikhail Rabinovich and Nikolai Verichev caused quite a stir in the scientific world. Here's how Valentine Afraimovich explained their work on synchronization in the system of interconnected stochastic oscillators in layman's terms:

– Famous 18th century physicist Christiaan Huygens has provided a perfect description for the synchronization of the periodic fluctuations. Two hundred years later, Soviet academician Aleksandr Andronov's experiments led him to conclude that non-periodic fluctuations sooner or later become synchronized as well. Our group has proven it through a mathematical model.

This discovery had an enormous practical impact, especially in the field of medicine. Non-periodic fluctuations are characteristic of the way the cells in the human brain function. When these cells start to move synchronously, it triggers an epileptic seizure. Understanding the "mathematics" of this process helped pharmacologists develop ways to prevent preliminary causes of an epileptic seizure and ease the suffering of the people afflicted by this condition.

Concurrently with his work at the Gorky's Scientific Research Institute, Valentine Afraimovich begun active cooperation with local physicists. In 1990, he and his frequent co-author Mikhail Rabinovich were invited to work in the US—first the University of San Diego, then Pennsylvania State University. Soon Valentine accepted a two-year position at the Georgia Institute of Technology in Atlanta, then the Northwestern University in Illinois. After that, he spent two years in East Asia, teaching graduate students and conducting research at the Tsing Hua University in Hsinchu City.

Valentine Afraimovich spent the next twenty years in Mexico, having received tenure as a research professor at the University of San Luis Potosi—Mexico's scientific and industrial hub. His works at the University of San Luis Potosi include famous collaborations with Albert Luo, Maurice Courbage, Edgardo Ugalde, Jose Urias, Mikhail Rabinovich, George Zaslavsky, Lev Glebsky.

The dawn of XXI century marks the beginning of Valentine Afraimovich's increasing interest in the workings of the human brain. He begun studying the dynamic mechanics of the autobiographical memory development, breaks in concentration, and nonlinear dynamics of consciousness as a hierarchical cognitive process. His extraordinary ability to think outside the box, to find fresh approach to seemingly unsolvable problems, and to use the beauty of the created model as a criterion of its potential for success have helped Valentine build the mathematical foundation for the dynamic theory of human intellectual activity. His results in this area will be featured in the book "The Mathematics of Consciousness", which Valentine Afraimovich began to write together with Pablo Varona (Madrid) and Mikhail Rabinovich (San Diego, CA).

Valentine Afraimovich was an exceptional teacher and played a pivotal role in the formation of many young scientists. He possessed encyclopedic knowledge of the theory of dynamic systems. Valentine's tremendous erudition and creative approach to any subject, distaste for generalizations, and ability to get to the core of the problem made his lectures fascinating and unforgettable for any audience. His other unique skill was the ability to boil down the most complex issues to their essentials and explain them with great precision and clarity. Any part of the world Valentine visited—be it Russia, the USA, Europe, China, Mexico—he nurtured followers, pupils and co-authors.

Valentine remained a lyric at heart. He devoured books, wrote poems, and was ever-ready to bang out a swinging tune on the piano. In high school he played piano in a light jazz trio. As a student at the university, Valentine toured the entire Soviet Union with a university jazz band "Nezabudka" ("Forget-me-nots"), where he met his future wife Ludmila. Invariably, he was the life of any party and managed to unite around him quite a diverse array of people. All his life he offered extensive support to his friends and colleagues and effortlessly left a warm, lingering memory of himself even after a brief encounter.

Until his last breath, Valentine Afraimovich was brimming with scientific and personal plans. He was working on new scientific publications, organizing a large conference at the university of San Luis Potosi, supervising doctorate applications of several of his students. He was very proud of the fact that, despite aging, his scientific mind continued to operate at full capacity. However, the scientist liked to joke that once the brain in his large, curly head becomes covered in cobwebs, he'll move in with his daughter Alexandra's family on the Pacific coast, where he'll swim, paint, and stir forgotten passions in the sunken bosoms of female retirees…

Alas, on January 20th, 2018, in the middle of a tango competition at the beautiful Mexican city of San Miguel de Allende, Valentine's wife Ludmila has suffered a ruptured aorta and died on the spot. Exactly one month later, on February 21st, 2018, stopped the heart of Valentine Afraimovich. The untimely death of this outstanding scientist and larger-than-life human being has orphaned both his family and the scientific community. His friends and colleagues, anyone who's ever met him, felt a gaping void in their hearts which was once filled by Valentine Afraimovich— infinitely kind, funny, and talented ray of light.

The Need for More Integration Between Machine Learning and Neuroscience

Adrián Hernández and José M. Amigó

Abstract Neuroscience and machine learning are two interrelated fields with obvious synergies. In recent years we have seen profound advances in these fields and in the integration between them. However, there remain challenges for a greater integration between the two disciplines. In this chapter we describe two areas, different but related, that can serve as a guide in the future for this integration. On the one hand, it is necessary to have more explanatory algorithms in order to understand information processing in the brain. The combination of multilayer and adaptive networks can be the right framework to understand this processing and analyse the interesting computational capabilities that occur in the brain. On the other hand, machine learning algorithms should have, similar to brain processing, more innate structure. This prior structure could make the process of learning more efficient and intuitive, and support artificial intelligence.

1 Introduction

In recent years we have seen profound advances in the fields of neuroscience and machine learning. The use of deep neural networks together with the parallel-processing capabilities of GPUs (Graphics Processing Unit) has allowed to improve the performance in problems of machine learning (image recognition, language processing, game playing, etc.) [1, 2]. In neuroscience, a revolution is taking place thanks to technologies such as optogenetics that allow neuronal control, and to open initiatives to share neural data [3–5].

Neuroscience and machine learning are two interrelated fields with obvious synergies [6–8]. Three traditional relationships between machine learning and neuroscience are the following.

The first is the use of machine learning techniques to analyze and predict data in neuroscience. Several projects, such as the connectome, are producing large

A. Hernández · J. M. Amigó (✉)
Centro de Investigación Operativa, Universidad Miguel Hernández, Av. de la Universidad s/n, 03202 Elche, Spain
e-mail: jm.amigo@umh.es

© Higher Education Press 2021
D. Volchenkov (ed.), *Nonlinear Dynamics, Chaos, and Complexity*,
Nonlinear Physical Science, https://doi.org/10.1007/978-981-15-9034-4_2

amounts of observations on the anatomical and functional connections of the brain [4]. Through the use of machine learning algorithms it is possible to classify, predict and extract conclusions from that wealth of data.

The second is the development of machine learning algorithms inspired by the brain. Artificial neural networks and deep learning emerged by emulating (and at the same time simplifying) the way in which the brain processes information [1, 9]. Evolution has made many biological processes an example of efficiency and robustness.

The third, which is a very important field of application, is the use of machine learning techniques to theorize and explain the functioning of the brain. The use of algorithms and ideas from machine learning can contribute in a prominent way to understand how information is processed in the brain, both at the biophysical and system level.

Despite the important developments in these three areas, a tighter integration between the two disciplines is necessary. The two disciplines must evolve together by adopting a common approach in which recent advances in machine learning can inspire new research in neuroscience and, the other way around, a better understanding of the brain could guide the development of new architectures, models and algorithms.

In this chapter we describe two areas, different but related, that can serve as a guide in the future for a greater integration between machine learning and neuroscience.

Recently, multilayer adaptive networks have been proposed as the appropriate framework for modelling neuronal processing [10, 11]. Multilayer because in the brain different synaptic and neuromodulatory layers interact to produce behaviour, and adaptive because the topology of these networks changes according to the dynamics. Within this framework, it is possible to identify and analyse interesting computational capabilities that occur in the brain, many of them similar to those used in machine learning.

Furthermore, the extra-synaptic (neuromodulatory) layers configure and specify the structure of neuronal circuits, adapting them to the environment and to the cognitive needs at each time.

In recent years, though, calls have been made for machine learning algorithms to assume more innate structure [12]. Only by assuming more innate structure, similar to what happens in brain processing, will it be possible to achieve more efficient and intuitive learning. This prior structure incorporates specific aspects of the target function.

2 Machine Learning and Neuroscience

Of the three traditional relationships between neuroscience and machine learning mentioned above, the use of machine learning to analyse data in neuroscience is a promising field with many challenges.

As pointed out in [13], the increasing data acquisition in neuroscience has created the need to leverage that information deluge to better understand how the brain works.

A large number of neuroscience databases have been compiled with information regarding neurons, gene expression, and microscopic and macroscopic brain structure. Although machine learning techniques have helped to analyze this data, much work remains to be done in this area [14].

In [15], the authors were able to define depression subtypes by clustering subjects according to patterns of functional connectivity. Even more, these biotypes were also informative in predicting which patients responded to a specific treatment.

However, in [16] the authors take a classical microprocessor as a model organism to see if data analysis techniques used in neuroscience can decode the way it processes information. They show that those techniques reveal interesting structure in the data but no relevant hierarchy of information processing. More importantly, the study suggests that availability of unlimited data is not sufficient to understand the functioning of the brain.

The design and development of bio-inspired machine learning algorithms is the second relationship between neuroscience and machine learning considered in the Introduction.

Machine learning has been influenced from the beginning by the biological processes that occur in the brain. For example, neural networks have become a key element in the development of machine learning in recent years.

Deep learning models use multiple levels of representation obtained by transforming the representation at one level into a representation at a higher, more abstract level [1]. These models try to emulate the activity in the neural layers of the neocortex, a part of the mammalian brain involved in higher-order brain functions.

Specific architectures of deep learning, such as convolutional neural networks [17], have also been inspired by biological processes. They replicate the hierarchical visual processing in the visual cortex. Furthermore, reinforcement learning, a technique inspired by behaviourist psychology in which an agent interacts with its environment to maximize some reward, has seen important progress in recent years with the use of deep neural networks [18].

Finally, the third relationship between neuroscience and machine learning that we mentioned before is the use of machine learning techniques to understand and model how information is processed in the brain. Computation, machine learning and applied mathematics are indeed a common source of interesting ideas and models for studying neuroscience.

Hodgkin-Huxley model [19] was a key achievement because it allowed to approximate the electrical characteristics of neurons and describe how action potentials in neurons are initiated and propagated. This model is relevant because it explains in a quantitative way some biological processes of the brain. However, it is necessary to go further and find algorithms that describe the underlying computation carried out in the brain. These models, similar to what is indicated in [13], should have some common characteristics:

(i) They must take into account all the biological components relevant to the process studied.
(ii) They must provide a description of how the observed processes and data are generated from the biological components.
(iii) A key aspect is to understand how the model works and how its components interact.
(vi) The model must be generalizable to similar processes in other subjects or organisms.

As we have seen, important advances have been made in the three traditional areas of relationship between machine learning and neuroscience. However, there remain challenges for a greater integration between the two disciplines. On the one hand it is necessary to have explanatory algorithms to better understand and model information processing in the brain. On the other hand, machine learning algorithms should have more innate structure, similar to brain processing. In the following sections we will see two promising approaches that try to address these challenges.

3 Multilayer Adaptive Networks in Neuronal Processing

The connectome is a wiring diagram mapping all the neural connections in the brain. It maps, at the cellular level, the neurons and synapses within a part or all of the brain and can be structural, if it describes anatomical connections, or functional, if it describes statistical associations.

The study of the connectome using graph theory has resulted in a significant advance in neuroscience [20–22]. Neurons are the nodes of the network and synapses correspond to the edges between those nodes. The use of new techniques for recording neural activity combined with network theory has allowed to link topological and dynamic patterns with mental disorders.

Modelling and analysing the connectome play a prominent role to understand neurotransmission (fast synaptic transmission) networks, in which neurotransmitters influence the postsynaptic neuron producing an inhibitory or excitatory response. However, knowledge and study of the connectome are not enough to understand how the brain processes information because neurons use other forms of communication as neuromodulation [23–25].

Neuromodulators modify neuronal circuit outputs by changing neuronal excitability and synaptic dynamics and strength [26]. They reconfigure the connectome and add new computational and processing capabilities to traditional synaptic transmission. Therefore, the connectome provides a minimal structure and neuromodulators shape the functional circuits that give rise to behaviour. Moreover, without taking into account these neuromodulatory layers, it is impossible to explain information processing in the brain.

Then, the question that arises is what models allow us to explain how the interaction of different layers (neurotransmission and neuromodulators) gives rise to information processing [11]?

In the last years, there has been a growing interest in adaptive and multilayer networks. Commonly, the study of dynamical networks has covered either dynamics on networks, in which nodes are dynamical systems coupled through static links, or dynamics of networks, where network topology evolves dynamically in time. But more recently adaptive networks have gained increasing interest [27–29]. In adaptive networks links change with the states of the nodes in an interplay between node states and network topology. An example of adaptive network is a neuronal network in which the firing rates and the synaptic connections interact with each other.

On the other hand, all the research has focused on single-layer networks, in which a single type of node is connected via a single type of link. But in most biological systems multiple entities interact with each other via multiple layers of connectivity, thus making it necessary to generalize network theory to study multilayer systems [30–32].

A typical multilayer network has the following components:

(i) A number N of nodes (denoted by Latin letters $i, j, \ldots$) and a number L of layers (denoted by Greek letters $\alpha, \beta, \ldots$).
(ii) Node $i \in \{1, 2, \ldots, N\}$ in layer $\alpha \in \{1, 2, \ldots, L\}$ has a state $s_{i\alpha}(t)$.
(iii) A 4th-order, time-dependent adjacency tensor $M(t)$ with components $m_{i\alpha}^{j\beta}(t)$ which are the weights of the link from any node i in layer α to any node j in layer β in the network. Self-links are excluded, so $m_{i\alpha}^{i\alpha}(t) = 0$.

In multilayer networks there can be different types of interactions between nodes: Intra-layer links within the same layer, inter-layer links between the same nodes in different layers and inter-layer links between different nodes in different layers. Multiplex networks, in which different layers are not interconnected except from each node to itself, are a special class of multilayer networks (see Fig. 1).

The combination of multilayer and adaptive networks is the right framework to understand neuronal information processing because:

(i) Neuromodulatory layers reconfigure the connectome by changing neuronal and synaptic properties.
(ii) These extra layers operate on temporal and spatial scales different from the fast synaptic ones and regulate neuronal processing.
(iii) The neuromodulatory layers interact with the neurotransmission layer in a complicated way, adding interesting computational capabilities.

In [33] the authors hold that further understanding of brain function and dysfunction will require an integrated framework, called "dynome", that links brain connectivity with brain dynamics. As pointed out in [34] for the crustacean stomatogastric nervous system, different regulatory mechanisms (synaptic and intrinsic neuronal properties, neuromodulation and gene expression regulation) influence each other to produce robustness and flexibility to circuit outputs. In [10] the C. elegans

Fig. 1 A multiplex network, with multiple layers of the same set of nodes and different types of relationship between them

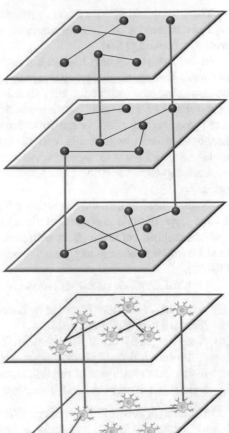

Fig. 2 A multilayer network with synaptic and neuromodulatory layers along with the corresponding communication links between layers

connectome is mapped as a multiplex network with synaptic, gap junction, and neuromodulatory layers representing alternative modes of interaction between neurons. The authors identified locations in the network where aminergic and neuropeptide signalling modulate synaptic activity.

We may conclude then that extending the connectome with synaptic and neuromodulatory layers seems the natural pathway to follow (see Fig. 2).

A simplified multilayer adaptive network model that accounts for these extralayers of interaction is defined in [11]. The first layer is a typical neurotransmission layer with adaptive node states and synapses. The remaining layers contain adaptive neuromodulatory states and links. The interaction between layers is of three types:

(i) Node states in the first layer are influenced by the states of the same node in the neuromodulatory layers.
(ii) Synapse weights in the first layer are influenced by the same link weights in the neuromodulatory layers.
(iii) Node states in the neuromodulatory layers are influenced by the state of the same node in the first layer.

Interesting insights can be extracted by examining basic aspects of this type of model. We can have extrinsic neuromodulation (i.e., neuromodulatory layers do not depend on neurotransmission activity) or intrinsic (i.e., neuromodulatory layers are not isolated from neurotransmission activity). Also, in most cases neuromodulation process is slower than the neurotransmission one.

Therefore, multilayer adaptive networks are a promising approach to take into account the different layers of neuronal communication and complete the connectome. Even more, they allow to analyze some computational capabilities added by neuromodulators such as circuit performance improvement [35], robustness [36, 37], circuit reconfiguration [38] and memory storage regulation.

4 More Innate Structure in Machine Learning Algorithms

The development of language is one of the most important cognitive processes in the human brain. Understanding how language is acquired and what biological preconditions are needed for language development are relevant subjects [39].

Traditionally, there have been two competing theories about language acquisition [40]:

(i) Nativist theory: Represented by Noam Chomsky, the theory argues that the learner possesses an innate capacity for dealing with language. The learner is biologically predisposed to learn languages. Chomsky says that all children have an innate language acquisition device (LAD).
(ii) Empiricist theory: According to the theory, there is enough information in the linguistic input and therefore there is no need to assume an innate structure. Learning begins with no knowledge of language but with the ability to learn it. In this model the determinant factor is the environment.

Although currently the language is not seen as the effect of the environment on a blank slate, there is a debate about the nature of the prior structure in language acquisition.

In a similar way, there is at present a constructive debate about the need of incorporating more structure into deep learning systems [41]. Assuming that a structure is necessary, the two points of view here differ in the role that this structure must have in the algorithm.

Possibly, the path towards artificial intelligence involves using more efficient algorithms that allow learning in a more intuitive way. For the following reasons, this is only possible by assuming more structure:

(i) As we have seen in the past, incorporating prior knowledge about the under-lying function has achieved great results. Convolutional neural networks [17] make the assumption that the processing of an image should be translationally invariant. The success of long short-term memories [42] comes from modelling temporal dependencies and structure, especially in language processing.

(ii) As pointed out in [12], humans and other animals are born with a lot of innate machinery.

(iii) The human brain develops its cognitive functions thanks to prior structures adapted to each function. Also, alterations in brain connectivity have been linked to mental disorders [43].

(iv) Only with more structure it is possible to design algorithms that can learn more from less data.

Then, how much structure is convenient to incorporate into algorithms in order to improve learning? According to [44], a good start is to understand the innate structure in humans minds and try to analyse if these mechanisms add value to artificial intelligence.

Also, in [12] the author proposes a list of innate machinery and structure to support artificial general intelligence (e.g., object representation, structured representations, causality, translational invariance, …).

Although the debate will continue, we must face it from an open perspective and use evolution and the human brain as a source of inspiration. Millions of years of human evolution have provided us with valuable structures and mechanisms for learning.

5 Conclusion and Outlook

Despite the relevant advances that have been made in machine learning, neuroscience and in the integration between both, there remain challenges for a greater integration.

In neuroscience it is fundamental to find algorithms that provide a description of how the observed processes and data are generated from the biological components. This algorithms must take into account all the biological components relevant to the process studied. Multilayer adaptive networks, in which different synaptic and neuromodulatory layers interact to produce behaviour, is a promising approach in that sense. They allow to analyse some interesting computational capabilities added by neuromodulators. Furthermore, in this models the extra-synaptic (neuromodulatory) layers configure and specify the structure of neuronal circuits, adapting them to the environment.

Regarding future research, we indicate next promising lines of application of multilayer adaptive networks:

(i) Concrete biological models of interesting phenomena via the multilayer approach could provide a description of the observed processes and how they are generated from the biological components.

(ii) Use of multilayer adaptive networks for the study of the computational capabilities added by the additional chemical layers (e.g., adaptability, regulation, robustness, degeneracy, memory, recurrency).

(iii) New theoretical and computational models, such as multilayer adaptive networks, will be required to explain new observations made with optogenetics. This technique has made it possible to differentiate between mechanisms of memory retrieval and memory storage [3, 45, 46].

There is currently a constructive debate about the need of more structure into deep learning systems. Probably, algorithms should have, similar to brain processing, more innate structure to learn in a more intuitive way. As seen in the past, incorporating prior knowledge has achieved good results, and in the future it can support artificial general intelligence. One way to obtain this innate machinery is to analyse what kind of mechanisms and structure have allowed the brain to perform such diverse functions so efficiently.

Looking forward, and as pointed out in [44], a possible roadmap to incorporate more prior knowledge in machine learning algorithms and support artificial intelligence is:

(i) For the different functions implemented by the brain, analyse what architectures, constraints and assumptions facilitate the learning process.

(ii) Try to incorporate this structure into algorithms that perform similar functions but in machine learning.

(iii) In general, use results on information processing in the brain to design new machine learning models.

With these two approaches, neuroscience and machine learning can benefit from each other, eliciting a positive contribution of mutual value:

(i) Advances in machine learning and mathematical modelling can be harnessed in neuroscience to describe the underlying computation carried out in the brain.

(ii) A better understanding of the brain, especially of its structure and mechanisms for information processing, could guide the development of new architectures, models and algorithms in machine learning.

This work was supported by the Spanish Ministry of Science and Innovation, grant PID2019-108654GB-I00.

References

1. LeCun Y, Bengio Y, Hinton G (2015) Deep learning. Nature 521:436–444
2. Yadan O, Adams K, Taigman Y, Ranzato M (2014) Multi-GPU training of ConvNets. arXiv:1312.5853

3. Tonegawa S, Pignatelli M, Roy DS, Ryan TJ (2015) Memory engram storage and retrieval. Curr Opin Neurobiol 35:101–109
4. Sporns O, Tononi G, Kötter R (2005) The human connectome: a structural description of the human brain. PLoS Comput Biol 1:e42
5. Horn A, Ostwald D, Reisert M, Blankenburg F (2014) The structural-functional connectome and the default mode network of the human brain. NeuroImage 102:142–151
6. Hinton G (2011) Machine learning for neuroscience. Neural Syst Circuits 1:12
7. Marblestone AH, Wayne G, Kording KP (2016) Toward an integration of deep learning and neuroscience. Front Comput Neurosci 10:94
8. Velde F (2010) Where artificial intelligence and neuroscience meet: the search for grounded architectures of cognition. Adv Artif Intell 2010:918062
9. McCulloch WS, Pitts W (1943) A logical calculus of the ideas immanent in nervous activity. Bull Math Biophys 5:115
10. Bentley B, Branicky R, Barnes CL, Chew YL, Yemini E, Bullmore ET, Vértes PE, Schafer WR (2016) The multilayer connectome of Caenorhabditis elegans. PLoS Comput Biol 12:e1005283
11. Hernández A, Amigó JM (2018) Multilayer adaptive networks in neuronal processing. Eur Phys J Spec Top 227:1039–1049. https://doi.org/10.1140/epjst/e2018-800037-y
12. Marcus G (2018) Innateness, AlphaZero, and artificial intelligence. arXiv:1801.05667
13. Vu MT, Adal T, Ba D, Buzsáki G, Carlson D, Heller K, Liston C, Rudin C, Sohal VS, Widge AS, Mayberg HS, Sapiro G, Dzirasa K (2018) A Shared vision for machine learning in neuroscience. J Neurosci 38:1601–1607
14. Helmstaedter M (2015) The mutual inspirations of machine learning and neuroscience. Neuron 86:25–28
15. Drysdale AT, Grosenick L, Downar J, Dunlop K, Mansouri F, Meng Y, Fetcho RN, Zebley B, Oathes DJ, Etkin A, Schatzberg AF, Sudheimer K, Keller J, Mayberg HS, Gunning FM, Alexopoulos GS, Fox MD, Pascual-Leone A, Voss HU, Casey BJ, Dubin MJ, Liston C (2017) Resting-state connectivity biomarkers define neurophysiological subtypes of depression. Nat Med 23:28–38
16. Jonas E, Kording KP (2017) Could a neuroscientist understand a microprocessor? PLoS Comput Biol 13:e1005268
17. Lecun Y, Bottou L, Bengio Y, Haffner P (1998) Gradient-based learning applied to document recognition. Proc IEEE 86:2278–2324
18. Arulkumaran K, Deisenroth MP, Brundage M, Bharath AA (2017) A brief survey of deep reinforcement learning. IEEE Signal Process Mag
19. Hodgkin AL, Huxley AF (1952) A quantitative description of membrane current and its application to conduction and excitation in nerve. J Physiol 117:500–544
20. Achard S, Bullmore E (2007) Efficiency and cost of economical brain functional networks. PLoS Comput Biol 3:e17
21. Baker ST, Lubman DI, Yücel M, Allen NB, Whittle S, Fulcher BD, Zalesky A, Fornito A (2015) Developmental changes in brain network hub connectivity in late adolescence. J Neurosci 35:9078–9087
22. Barnett L, Buckley CL, Bullock S (2009) Neural complexity and structural connectivity. Phys Rev E 79:051914
23. Bargmann C, Marder E (2013) From the connectome to brain function. Nat Methods 10:483–490
24. Bargmann C (2012) Beyond the connectome: how neuromodulators shape neural circuits. BioEssays 34:458–465
25. Brezina V (2010) Beyond the wiring diagram: signalling through complex neuromodulator networks. Philos Trans R Soc B 365:2363–2374
26. Nadim F, Bucher D (2014) Neuromodulation of neurons and synapses. Curr Opin Neurobiol 29:48–56
27. Sayama H, Pestov I, Schmidt J, Bush BJ, Wong C, Yamanoi J, Gross T (2013) Modeling complex systems with adaptive networks. Comput Math Appl 65:1645–1664

28. Maslennikov OV, Nekorkin VI (2017) Adaptive dynamical networks. Physics-Uspekhi 60:694–704
29. Aoki T, Rocha LEC, Gross T (2016) Temporal and structural heterogeneities emerging in adaptive temporal networks. Phys Rev E 93:040301
30. Kivelä M, Arenas A, Barthelemy M, Gleeson JP, Moreno Y, Porter MA (2014) Multilayer networks. J Complex Netw 2:203–271
31. De Domenico M, Granell C, Porter MA, Arenas A (2016) The physics of spreading processes in multilayer networks. Nat Phys 12:901–906
32. De Domenico M (2017) Multilayer modeling and analysis of human brain networks. Gigascience 6:1–8
33. Kopell N, Gritton HJ, Whittington MA, Kramer MA (2014) Beyond the connectome: the dynome. Neuron 83:1319–1328
34. Daur N, Nadim F, Bucher D (2016) The complexity of small circuits: the stomatogastric nervous system. Curr Opin Neurobiol 41:1–7
35. Holca-Lamarre R, Lücke J, Obermayer K (2017) Models of acetylcholine and dopamine signals differentially improve neural representations. Front Comput Neurosci 11:54
36. O'Leary T, Williams AH, Caplan JS, Marder E (2013) Correlations in ion channel expression emerge from homeostatic tuning rules. PNAS 110:E2645–54
37. Marder E, Goeritz ML, Otopalik AG (2015) Robust circuit rhythms in small circuits arise from variable circuit components and mechanisms. Curr Opin Neurobiol 31:156–163
38. Gutierrez GJ, Marder E (2014) Modulation of a single neuron has state-dependent actions on circuit dynamics. eNeuro 1. ENEURO.0009-14.2014
39. Mueller JL, Männel C, Friederici AD (2015) Biological preconditions for language development. In: Wright JD (ed) International encyclopedia of the social & behavioral sciences, 2nd edn. Elsevier Press, Oxford, pp 650–655
40. da Silva RB (2007) A brief discussion on the biological factors in the acquisition of language. Revista do GEL S J do Rio Preto 4:153–169
41. Deep Learning, Structure and Innate Priors. A Discussion between Yann LeCun and Christopher Manning. http://www.abigailsee.com/2018/02/21/deep-learning-structure-and-innate-priors.html
42. Hochreiter S, Schmidhuber J (1997) Long short-term memory. Neural Comput 9:1735–1780
43. Zeng K, Kang J, Ouyang G, Li J, Han J, Wang Y, Sokhadze EM, Casanova MF, Li X (2017) Disrupted brain network in children with autism spectrum. Sci Rep 7:16253
44. Marcus G (2018) Deep learning: a critical appraisal. arXiv:1801.00631
45. Ryan TJ, Roy DS, Pignatelli M, Arons A, Tonegawa S (2015) Memory. Engram cells retain memory under retrograde amnesia. Science 348:1007–1013
46. Titley HK, Brunel N, Hansel C (2017) Toward a neurocentric view of learning. Neuron 95:19–32

Quasiperiodic Route to Transient Chaos in Vibroimpact System

Victor Bazhenov, Olga Pogorelova, and Tatiana Postnikova

Abstract Dynamic behaviour of the nonsmooth systems is interesting and explored subject in nonlinear science. We studied quasiperiodic route to chaos in nonlinear nonsmooth discontinuous vibroimpact system. In narrow frequency range different oscillatory regimes have succeeded each other many times under very small control parameter varying. There were periodic subharmonic regimes—chatters, quasiperiodic, and chaotic regimes. There were the zones of transition from one regime to another, the zones of prechaotic or postchaotic motion. The hysteresis effects (jump phenomena) occurred for increasing and decreasing frequencies. The chaoticity of obtained regime has been confirmed by typical views of Poincaré map and Fourier spectrum, by the positive value of the largest Lyapunov exponent, and by the fractal structure of Poincaré map. Discontinuous bifurcations are also described—it is phenomenon unique for nonsmooth systems. It is discussed the largest Lyapunov estimation for nonsmooth vibroimpact system.

Dedication

It is a big honour for us to publish the chapter in the book devoted to the memory of Professor Valentin Afraimovich. He was the Great Scientist and the Great Man. His works on nonlinear dynamics in general and about the chaos and transient chaos in particular are helpful to guide the one interested in contemporary science. To say more they are providing a lot of exciting ideas.

V. Bazhenov (✉) · O. Pogorelova · T. Postnikova
Kyiv National University of Construction and Architecture,
31, Povitroflotskiy avenu, Kyiv, Ukraine
e-mail: vikabazh@ukr.net

O. Pogorelova
e-mail: pogos13@ukr.net

T. Postnikova
e-mail: posttan@ukr.net

© Higher Education Press 2021
D. Volchenkov (ed.), *Nonlinear Dynamics, Chaos, and Complexity*,
Nonlinear Physical Science, https://doi.org/10.1007/978-981-15-9034-4_3

1 Introduction

At present chaotic vibrations are the one of the most interesting and investigated subjects in nonlinear dynamics. There are many papers, monographs and textbooks about dynamic behaviour in general and routes to chaos in particular in smooth nonlinear systems. Also there are many works devoted to studies of nonsmooth discontinuous dynamical systems, for example [1–9]. But dynamical processes in nonsmooth systems are studied less. In work [5] authors divide nonsmooth discontinuous dynamical systems into three types according to their degree of discontinuity. There are among them Fillipov systems and the impacting systems with velocity reversals. Moreover the systems with impacts between its elements have the grossest form of nonlinearity and the non-smoothness. Many new phenomena unique to nonsmooth systems are observed under variation of system parameters. Recently the investigations of such systems are developed rapidly. Especially systems with impacts are of the particular interest for scientists. Exactly in such systems the discontinuous dangerous bifurcations are arising under system parameters variation. Just such hard bifurcations can portend the crisis and catastrophe.

Vibroimpact system is strongly nonlinear nonsmooth one; the set of differential equations of motion contains the discontinuous right-hand side. We have observed phenomena unique to nonsmooth systems—the discontinuous bifurcations under the changing of external force amplitude and frequency. At points of discontinuous bifurcations Floquet multiplier is leaving the unit circle by jump. Sometimes this jump may be very big one.

The studying of vibroimpact system dynamic behaviour both in general and for concrete system is of the special interest. In particular the routes to chaos in such systems also are of the special interest. It is well known that completely deterministic dynamic system may begin to behave by unforeseen chaotic manner when any accidental influence is absent. However, in this unpredictability it is possible to identify a number of regularities in the system behaviour which distinguishes this phenomenon from the classical random processes. Moreover, in contrast to the classical random processes, the phenomenon of deterministic chaos can be reproduced in natural, laboratory and numerical experiments. Just deterministic chaos is not an exceptional mode of dynamical systems behaviour; on the contrary, such regimes are observed in many dynamical systems in mathematics, physics, chemistry, biology and medicine. Recently such phenomena are more often described in studies on economics, sociology, philosophy, history [10]. Therefore, the studying of chaotic dynamics is one of the main ways of modern natural science development. Many monographs, papers and textbooks are devoted to chaos studying [11–16].

It is known three main routes to chaos in dynamical systems [11, 12]:

- period-doubling route to chaos—the most celebrated scenario for chaotic vibrations;
- quasiperiodic route to chaos;

- route to chaos via intermittency by Pomeau and Manneville: the long periods of periodic motion with bursts of chaos; as one varies a parameter the chaotic bursts become more frequent and longer [17].

We use different characteristics in order to be sure that obtained oscillatory regime is chaotic one.

Poincaré maps are one of the principal ways of recognizing chaotic vibrations in low-degree of freedom problems. Poincaré maps and phase plane portraits can often provide graphic evidence for chaotic behaviour and the fractal properties of strange attractors. Poincaré maps help to distinguish between various qualitative states of motion such as periodic, quasiperiodic, or chaotic. But quantitative measures of chaotic dynamics are also important and in many cases are the only hard evidence for chaos.One of the significant characteristics is Fourier distribution of frequency spectra. The difference between chaotic and quasiperiodic motion can be detected by taking the Fourier spectrum of the signal. A quasiperiodic motion will have the well-pronounced peaks at basic frequencies and at their combinations, chaotic motion—a broad continuous spectrum of Fourier components.

Chaos in dynamics implies a sensitivity of the outcome of a dynamical process to changes in initial conditions. Small uncertainties in initial conditions lead to divergent orbits in the phase space. Small changes in initial conditions (or in some other parameters such as, for example, the amplitude or frequency of exciting force, damping coefficient) can dramatically change the type of output from a dynamical system.

Lyapunov exponents characterize the kind of dynamical system motion because they measure the divergence rate of nearby trajectories. In order to have a criterion for chaos one need only calculate the largest exponent λ which tells whether nearby trajectories diverge ($\lambda > 0$) or converge ($\lambda < 0$) on the average. Its sign is chaos criterion. For regular motions $\lambda \leq 0$, but for chaotic motion $\lambda > 0$ that is positive Lyapunov exponent imply chaotic dynamics.

There are some difficulties with Lyapunov exponents determination for non-smooth systems especially for discontinuous systems with impacts. These difficulties are caused by the discontinuity of motion equations right-hand sides. The Jacobian matrix which is used in well known Benettin's algorithm of Lyapunov exponent calculation [11, 12, 18] is also discontinuous. Therefore we have tried to estimate the largest Lyapunov exponent by numerical integration two copies of dynamic system with nearby initial conditions and following the distance evolution between them and have used three different formulas for such calculation. Besides we have succeeded to use Benettin's algorithm and have compared the results of these estimations.

At present there are several propositions for Lyapunov exponents calculation in nonsmooth systems. The authors of these propositions describe their own methods for such estimation [19–23].

One can consider the fractal structure of Poincaré map as visit card of chaotic motion. When the motion is chaotic, a mazelike, multisheeted structure in section may appear. This threadlike collection of points seems to have further structure when examined on a finer scale. The term *fractal* characterizes such Poincaré patterns. So the fractal dimension of chaotic attractor is one of the principal measures of chaos.

In [12] the author advises not to rely on one measure of chaos in dynamical experiments, but to use two or more techniques such as Poincaré maps, Fourier spectra, Lyapunov exponents or fractal dimension measurements before pronouncing a system chaotic or strange.

"Since quasi-periodic oscillations are ubiquitous in this real world full of various rhythms, bifurcations of quasi-periodic tori have been intensively studied" [24]. The birth and breakdown of invariant tori have been studied in such works as [25–28]. When we say about quasiperiodic route to chaos we have to bear in mind that in this case the whole picture is found sufficiently complicated. Its many aspects remain not studied to the end so far. Attractor evolution under governing parameter changing may be various and complicated. Quasiperiodic and periodic regimes may alternate and undergo different bifurcations.

After the breakup of torus we observed periodic subharmonic regimes—chatters, quasiperiodic, and chaotic regimes. There were the zones of transition from one regime to another, the zones of prechaotic or postchaotic motion. We think that observed chaotic motions in narrow frequency ranges may be considered as transient chaos [12] because chaotic vibrations appear for some parameter changes and then degenerate into a quasiperiodic motion after a short time. Such a chaotic set does not attract trajectories from its neighbourhood, it is nonattracting one.

2 The Reason for Studying of Quasiperiodic Route to Chaos

2.1 The Initial Equations. Loading Curves and Amplitude-Frequency Response

We have studied the dynamic behaviour of 2-DOF two-body vibroimpact system (Fig. 1) in our previous works [29–31].

This model is formed by the main body m_1 and attached one m_2, which can play the role of percussive or non-percussive dynamic damper. Bodies are connected by linear elastic springs with stiffness k_1 and k_2 and dampers with damping coefficients c_1 and c_2. (The damping force is taken as proportional to first degree of velocity with coefficients c_1 and c_2.)

Fig. 1 Vibroimpact system model

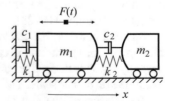

The differential equations of its movement are:

$$\ddot{x}_1 = -2\xi_1\omega_1\dot{x}_1 - \omega_1^2 x_1 - 2\xi_2\omega_2\chi(\dot{x}_1 - \dot{x}_2) -$$
$$- \omega_2^2\chi(x_1 - x_2 + D) + \frac{1}{m_1}[F(t) - F_{con}(x_1 - x_2)], \tag{1}$$
$$\ddot{x}_2 = -2\xi_2\omega_2(\dot{x}_2 - \dot{x}_1) - \omega_2^2(x_2 - x_1 - D) + \frac{1}{m_2}F_{con}(x_1 - x_2),$$

where $\omega_1 = \sqrt{\dfrac{k_1}{m_1}}, \omega_2 = \sqrt{\dfrac{k_2}{m_2}}; \xi_1 = \dfrac{c_1}{2m_1\omega_1}, \xi_2 = \dfrac{c_2}{2m_2\omega_2}; \chi = \dfrac{m_2}{m_1}.$

External loading is periodic one: $F(t) = P\cos(\omega t + \varphi_0), T = \dfrac{2\pi}{\omega}$ is its period.

Impact is simulated by contact interaction force F_{con} according to contact quasistatic Hertz's law:

$$F_{con}(z) = K[H(z)z(t)]^{3/2},$$
$$K = \frac{4}{3}\frac{q}{(\delta_1 + \delta_2)\sqrt{A+B}}, \delta_1 = \frac{1 - \nu_1^2}{E_1\pi}, \delta_2 = \frac{1 - \nu_2^2}{E_2\pi}, \tag{2}$$

where $z(t)$ is the relative closing in of bodies, $z(t) = x_2 - x_1, A, B$, and q are constants characterizing the local geometry of the contact zone; ν_i and E_i are respectively Poisson's ratios and Young's modulus for both bodies, $H(z)$ is the discontinuous step Heviside function. The numerical parameters of this system are following:

$m_1 = 1000\,\text{kg}, \omega_1 = 6.283\,\text{rad} \cdot \text{s}^{-1}, \quad \xi_1 = 0.036, E_1 = 2.1 \cdot 10^{11}\,\text{N} \cdot \text{m}^2, \quad \nu_1 = 0.3,$

$m_2 = 100\,\text{kg}, \omega_2 = 5.646\,\text{rad} \cdot \text{s}^{-1}, \quad \xi_2 = 0.036, E_2 = 2.1 \cdot 10^{11}\,\text{N} \cdot \text{m}^2, \quad \nu_2 = 0.3,$

$P = 500\,\text{N}, A = B = 0.5\,\text{m}^{-1}, \quad q = 0.318.$

We have obtained loading curves [29] and amplitude-frequency response [30] in wide range of control parameter by parameter continuation method. Periodic motion stability or instability was determined by matrix monodrómy eigenvalues that is by Floquet multipliers' values. The periodical solution is becoming unstable one if even though one Floquet multiplier leaves the unit circle in complex plane that is its modulus becoming more than unit.

The loading curves give the oscillation semi-amplitude dependences on excitation amplitude for both system bodies. Their global view in wide range of excitation amplitude is given at Fig. 2a. Excitation frequency is $\omega = 7.23\,\text{rad} \cdot \text{s}^{-1}$.

Here and further the upper curve corresponds to attached body (m_2), the lower one to main body (m_1). Unstable regimes are dotted by grey colour. At axis of ordinates we have semi-amplitude A_{max}. It means half the peak-to-peak amplitude. For the nonharmonic oscillation it is calculated by the formula $A_{max} = \dfrac{|x_{max}| + |x_{min}|}{2}.$

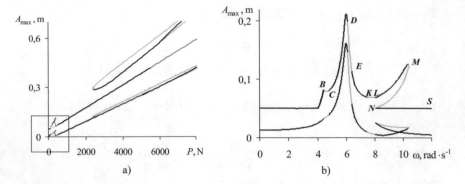

Fig. 2 **a** Loading curves in wide range of excitation amplitude; **b** Amplitude-frequency response in wide range of excitation frequency

The amplitude-frequency responses give the oscillation semi-amplitude dependence on excitation frequency for both system bodies. Their global view in wide range of excitation frequency is given at Fig. 2b. Excitation amplitude is $P = 500 \, \text{N}$.

Under frequency changing we have observed hysteresis effect[1] at the region *LMNS*. There are some regions of instability of basic $(1,1)$-regime[2] (T-periodic regime with 1 impact per cycle): *BC, DE, KL, MN*.

2.2 Neimark-Sacker Bifurcations

Now we'll pay attention at dynamic behaviour of vibroimpact system in frequency range $7.45 \, \text{rad} \cdot \text{s}^{-1} < \omega < 8.0 \, \text{rad} \cdot \text{s}^{-1}$ that is at the region *KL*. At points K and L stable $(1,1)$-regime is losing stability, the quasiperiodic regimes are arising as a result of Neimark-Sacker bifurcations. The two complex conjugate multipliers μ and μ^* are leaving the unit circle (Fig. 3).

After the stability loss at bifurcation points K and L the second basic oscillatory frequency is arising: $\omega_1 = \frac{1}{T}(arg\mu + 2k\pi), k = 0, \pm1, \pm2, \ldots$ (the argument of complex number is determined with accuracy $\pm 2k\pi$) [11]. This frequency ω_1 is not commensurate with first basic frequency ω. So the branching dynamical state is quasiperiodic one. Simultaneous time trace of phase plane motion and Poincaré map of this regime for $\omega = 7.46 \, \text{rad} \cdot \text{s}^{-1}$ are depicted at Fig. 4. Here and further (except Figs. 17 and 18) phase trajectories and Poincaré maps are presented for the main body m_1.

Its Fourier spectrum in logarithmic scale is also shown at Fig. 4. We see Poincaré section to be closed curve, and Fourier spectrum has the well-pronounced peaks

---

[1] We consider hysteresis effect as dependence of the system state on its history (the system manifests hysteretic features in the transition between different types of motion) [12, 16].

[2] The mark (n,k) means nT-periodic vibration with k impacts per cycle [32], T is period of external loading $T = 2\pi/\omega$.

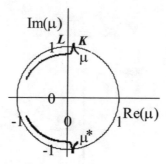

Fig. 3 Multipliers behaviour at KL region

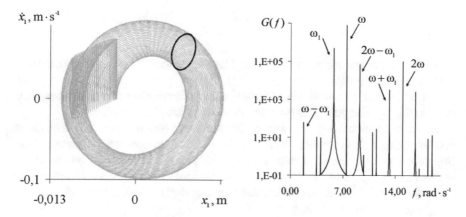

Fig. 4 Poincaré map and Fourier spectrum for quasiperiodic regime

at two basic frequencies ω and ω_1 and at their combinations, what is typical for quasiperiodic motion.

What does happen with system motion between points K ($\omega = 7.45\,\mathrm{rad} \cdot \mathrm{s}^{-1}$) and L ($\omega = 8.0\,\mathrm{rad} \cdot \mathrm{s}^{-1}$) ? What regimes do exist at this frequency range ?

These questions will be discussed in details at Sect. 5.

3 Discontinuous Bifurcations

3.1 Discontinuous Bifurcations Under Excitation Amplitude Varying

Under loading curve construction we observe the phenomenon unique for non-smooth system with discontinuous right-hand side — discontinuous bifurcation where stable impactless regime is becoming the unstable periodic one with impacts. Other

Fig. 5 Partial view of loading curves

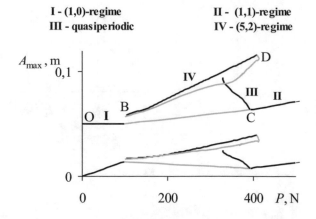

I - (1,0)-regime
III - quasiperiodic

II - (1,1)-regime
IV - (5,2)-regime

periodic regimes are arising at this point. Here set-valued Floquet multipliers cross the unit circle by jump.

Let us look at partial view of loading curves where discontinuous bifurcation occurs (Fig. 5). The point B is the point of discontinuous bifurcation.

It is phenomenon unique for nonsmooth nonlinear system whose equations have discontinuous right-hand side. The vibroimpact system converts its motion from impactless one (region OB) into motion with periodic impacts—unstable (1,1)-regime. Other regimes—stable and unstable branches of (5,2)-regime—are arising here. At point B Floquet multipliers are experiencing a discontinuous change and accepting big values [4, 5, 33].

The set-valued Floquet multipliers cross the unit circle in direction of real positive axis by jump that is their moduli becoming more than unit by jump. Floquet multipliers behaviour in the excitation amplitude range $0 < P < 500\,\text{N}$ is shown at Fig. 6.

Table 1 shows these jumps by numbers.

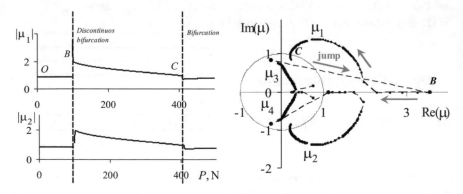

Fig. 6 Floquet multipliers behaviour in OBC region and its jumps under discontinuous bifurcation

Table 1 Floquet multipliers μ_1 and μ_2 jumps at point B

P	$\mathrm{Re}\mu_1$	$\mathrm{Im}\mu_1$	$|\mu_1|$	$\mathrm{Re}\mu_2$	$\mathrm{Im}\mu_2$	$|\mu_2|$
98.48	−0.25	0.83	0.87	−0.25	0.83	0.87
98.98	−0.25	0.83	0.87	−0.25	0.83	0.87
99.98	3.57	0	3.57	1.12	0	1.12

3.2 Discontinuous Bifurcations Under Excitation Frequency Varying

Under excitation frequency changing discontinuous bifurcations also occur. Let us look at partial views of amplitude-frequency response in narrow frequency ranges where discontinuous bifurcations occur (Fig. 7).

At point B we again observe phenomenon unique for nonsmooth systems with discontinuous right-hand side. This point B is the point of discontinuous bifurca-

Fig. 7 Partial views of frequency-amplitude response

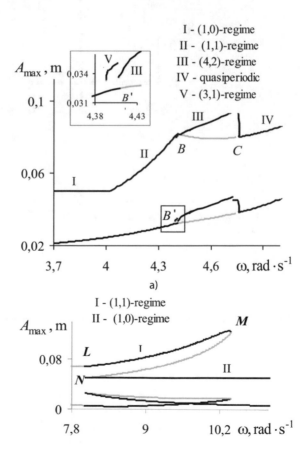

tion. The vibroimpact system converts its motion from impactless one into motion with periodic impacts. T-periodic stable impactless regime is becoming T-periodic unstable regime with one impact per cycle—(1,1)-regime. Other regimes are arising here—stable (3,1)-regime and stable (4,2)-regime. Let us note by the way that (3,1)-periodic regime is stable in small frequency range. It is rare attractor [34].

At point B two complex conjugate Floquet multipliers μ_1 and μ_2 are leaving the unit circle. They are experiencing change by jump and accepting big values (Fig. 8). It is obvious that one can also see the jump of monodromy matrix at this point. Monodromy matrix and multipliers values under discontinuous bifurcation are shown lower.

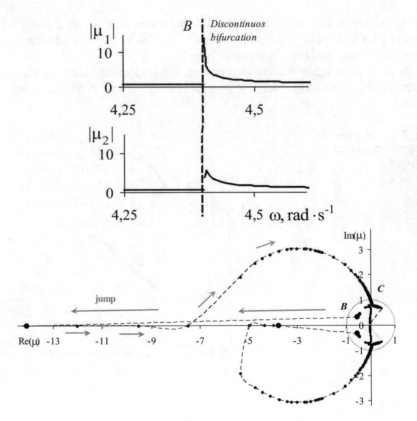

Fig. 8 Set-valued Floquet multipliers jumps under discontinuous bifurcation

Exitation frequency $\omega = 4.4041$ rad $\cdot$ s^{-1}

$$B_1 = \begin{vmatrix} 0.20 & 1.46 & 0.08 & 0.08 \\ 0.15 & 0.31 & 0.03 & 0.03 \\ -1.81 & -7.94 & 0.18 & 0.18 \\ 0.79 & 0.77 & 0.14 & 0.14 \end{vmatrix} \quad \begin{aligned} \mu_1 &= 0.45 + i0.66 \\ \mu_2 &= 0.45 - i0.66 \\ \mu_3 &= -0.57 + i0.33 \\ \mu_4 &= -0.57 - i0.33 \end{aligned}$$

Exitation frequency $\omega = 4.4042$ rad $\cdot$ s^{-1}

$$B_1 = \begin{vmatrix} -40.56 & -23.71 & -3.12 & 14.49 \\ 0.03 & 0.37 & 0.02 & 0.04 \\ -19.11 & -18.82 & -1.20 & 7.42 \\ -70.85 & -42.4 & -5.37 & 24.18 \end{vmatrix} \quad \begin{aligned} \mu_1 &= -14.08 + i0 \\ \mu_2 &= -0.77 - i0 \\ \mu_3 &= -0.05 - i0.06 \\ \mu_4 &= -0.57 - i0.33 \end{aligned}$$

Let us look now at Fig. 7b. At turning point M Floquet multiplier μ_1 passes through the unit circle over the $+1$ and moves along the positive real axis far away. Its way under MN unstable regime is shown at Fig. 9. Its velocity in motion along the axis is increasing and we see its appearance less and less often. The multiplier motion along positive real axis is demonstrated by Table 2.

At point N we again observe phenomenon unique to discontinuous system—discontinuous fold bifurcation. Here a stable branch and an unstable branch are connected. Set-valued Floquet multiplier μ_1 is returning (or is leaving) into the unit circle by huge jump (Fig. 9).

The abrupt jump of multiplier μ_1 is seen very well at Fig. 10.

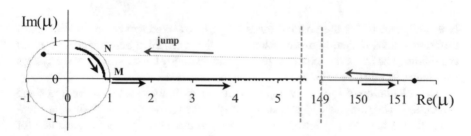

Fig. 9 Floquet multiplier μ_1 motion at MN region and huge jump at point N under discontinuous bifurcation

Table 2 Floquet multipliers μ_1 changing and jump at point N of fold discontinuous bifurcation

ω_i, rad $\cdot$ s^{-1}	8.03	8.04	8.5	8.06	8.07	8.10	8.16
Reμ_1	0.595	0.593	151.4	90.1	65.9	37.4	19.7
Imμ_1	0.654	0.652	0	0	0	0	0
$|\mu_1|$	0.8828	0.8829	151.4	90.1	65.9	37.4	19.7

Fig. 10 Discontinuous fold
bifurcation

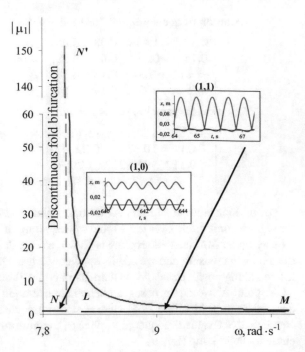

4 Lyapunov Exponents

It is well known that the positive largest Lyapunov exponent is one of the chaos criterion. Chaos in deterministic systems implies a sensitive dependence on initial conditions. The Lyapunov exponent test measures exactly the sensitivity of the system to changes in initial conditions.

The calculation of Lyapunov exponent in discontinuous system is finding some difficulties. They are caused just discontinuity of the right-hand side of motion equations. As we have said in Introduction at present there are several propositions for Lyapunov exponents calculation for nonsmooth systems where the authors describe their methods for such estimation [19–23].

4.1 The Largest Lyapunov Exponent Calculation

Conceptually, one imagines a small ball of initial conditions in phase space and looks at its deformation into ellipsoid under the dynamics of the system. If d is the maximum length of the ellipsoid, and d_0 is the initial size of the initial condition sphere, the Lyapunov exponent λ is interpreted by the equation

$$d(t) = d_0 e^{\lambda(t-t_0)} \tag{3}$$

Here we can consider d_0 as a measure of the initial distance between the two starting points, in a small but later time the distance becomes d.

In other words Lyapunov exponents for dynamic system with continuous time define the degree of distance or rapprochement for different but nearby trajectories at infinity that is the exponentially fast divergence or convergence of nearby orbits in phase space. This means that if two trajectories start close to one another in phase space, they will move exponentially away from each other for small times on the average.

If largest Lyapunov exponent is positive then the distance between initially nearby trajectories is increasing in the course of time. If it is negative then near trajectories is approaching each other some more, if it is zero then near trajectories are staying at the same distance approximately. Let us note that Lyapunov exponents may be different ones for different initial values. So one measurement is not sufficient and the calculation must be averaged over different regions of phase space. This average can be represented by

$$\lambda = \lim_{N \to \infty} \frac{1}{N} \sum_{i=1}^{N} \frac{1}{t_i - t_{0i}} \ln \frac{d_i}{d_{0i}} \tag{4}$$

Lyapunov exponents may be obtained analytically extremely rarely. There are numerical methods which allow their obtaining with acceptable accuracy.

The calculation of largest Lyapunov exponent is especially important for diagnostic of complicated dynamics regimes. More than that it is enough very often to know the sign of largest Lyapunov exponent—the presence of positive largest exponent is one of the chaos criterion. There are three possibility for such calculation:

- we numerically integrate two copies of dynamic system with nearby initial conditions and follow the distance evolution between them;
- we jointly numerically integrate the main equations and equations in the variations;
- we determine Lyapunov exponents from time series.

The first method may be used when we have got some difficulties with obtaining or numerical solving the equation in variations [11, 35, 36]. The second method is the most used. There is famous algorithm by Benettin and all [11, 12, 18] and there is the special software for its realization. The third method is used when we have not the dynamical equations and must estimate Lyapunov exponents from an experimental time series [37].

At first we have calculated the largest Lyapunov exponent by the first method that is by numerical integration two copies of dynamic system with nearby initial conditions and following the distance evolution between them [38].

The right-hand side of differential equations of motion contains the step Heviside function which is discontinuous one:

$$H(z) = \begin{cases} 1, z \geq 0 \\ 0, z < 0 \end{cases} \tag{5}$$

Therefore its differentiation has got some difficulties. When we use the first method for the largest Lyapunov exponent calculation we integrate the motion equations Eq. (1) with two nearby initial conditions. Let us note by the way that we integrate these equations by the program ode23s (MATLAB ®ODE solvers). This program integrates the systems of stiff differential equations. So we obtain two nearby trajectories. We must follow their distance evolution. It may be fulfilled by the different three formulas. We introduce the following notations.

The coordinates of representing point in phase space for the first and the second trajectories are setting by the vectors $\mathbf{X}(t)$, $\mathbf{Y}(t)$ accordingly:

$$\mathbf{X}(x_1(t), x_2(t), \dot{x}_1(t), \dot{x}_2(t), x_3(t)). \tag{6}$$

The five coordinate is connected with transition from nonautonomous problem to autonomous one and introducing the new variable $x_3(t) = \omega \cdot t$.

The initial conditions for two trajectories are $\mathbf{X}(0)$, $\mathbf{Y}(0)$.

Three formulas for largest Lyapunov exponent λ are as follows:

- We consider one trajectory piece and observe the changing of distance from d_0 to d between two trajectories. Then

$$\lambda \approx \frac{1}{T} \ln \frac{d}{d_0}, \text{ where } d = \|\mathbf{X}(T) - \mathbf{Y}(T)\|, \quad d_0 = \|\mathbf{X}(0) - \mathbf{Y}(0)\|, \tag{7}$$

T—the trajectory piece length in time.
- We consider M trajectory pieces of the time length T and compare the distance between two trajectories at beginning and at the end of each piece. Then

$$\lambda \approx \frac{1}{MT} \sum_{i=1}^{M} \ln \frac{d_i}{d_{i-1}}, \text{ where } d_i = \|\mathbf{X}(iT) - \mathbf{Y}(iT)\|. \tag{8}$$

- We consider M trajectory pieces of the same time length T and compare the distance between two trajectories at the end of each piece with initial distance. Then

$$\lambda \approx \frac{1}{MT} \sum_{i=1}^{M} \ln \frac{d_i}{d_0}, \text{ where } d_i = \|\mathbf{X}(iT) - \mathbf{Y}(iT)\|, \quad d_0 = \|\mathbf{X}(0) - \mathbf{Y}(0)\|. \tag{9}$$

Then in spite of difficulties under obtaining and numerical solving the equations in variations we have succeeded in using the Benettin's algorithm for the largest Lyapunov exponent calculation. So we had the possibility to compare the obtained results. This algorithm is described in all textbooks, so we will not repeat its description. Note only the following.

When we calculate the largest Lyapunov exponent by Benettin's algorithm we follow the evolution of variations vector

$$\mathbf{X}(\tilde{x}_1(t), \tilde{x}_2(t), \dot{\tilde{x}}_1(t), \dot{\tilde{x}}_2(t), \tilde{x}_3(t)).$$

The largest Lyapunov exponent is given as

$$\lambda \approx \frac{1}{MT} \sum_{i=1}^{M} \ln \|\tilde{x}_i\| . \tag{10}$$

If the differential equations of motion Eq. (1) are described in vector form as

$$\dot{x} = X(x), \tag{11}$$

then evolution of small excitement $\tilde{x}_i$ in linear approach is described by the equation

$$\dot{\tilde{x}} = J(t)\tilde{x}. \tag{12}$$

The matrix of equations in variations J (Jacobian matrix) for our vibroimpact system (Fig. 1 and formulas (1), (2)) have got the form:

$$J = \left(\begin{array}{cccc|c} 0 & 0 & 1 & 0 & 0 \\ 0 & 0 & 0 & 1 & 0 \\ a_{31} & a_{32} & a_{33} & a_{34} & a_{35} \\ a_{41} & a_{42} & a_{43} & a_{44} & 0 \\ 0 & 0 & 0 & 0 & 0 \end{array} \right) \tag{13}$$

Here

$$a_{31} = -\omega_1^2 - \omega_2^2 \chi - \frac{1}{m_1} \frac{\partial F_{con}}{\partial x_1}, \quad a_{32} = -\omega_2^2 - \omega_2^2 \chi - \frac{1}{m_2} \frac{\partial F_{con}}{\partial x_2},$$

$$a_{33} = -2\xi_1 \omega_1 - 2\xi_2 \omega_2 \chi, \quad a_{34} = 2\xi_2 \omega_2, \quad a_{35} = -\frac{P}{m_1} \sin(x_3 + \varphi_0),$$

$$a_{41} = \omega_2^2 + \frac{1}{m_2} \frac{\partial F_{con}}{\partial x_1}, \quad a_{42} = -\omega_2^2 + \frac{1}{m_2} \frac{\partial F_{con}}{\partial x_2}, \quad a_{43} = 2\xi_2 \omega_2, \quad a_{44} = -2\xi_2 \omega_2.$$

According to formula (2) $F_{con}(x_1 - x_2) = K \cdot (x_1 - x_2)^{\frac{3}{2}} \cdot H(x_1 - x_2)$. Then

$$\frac{\partial F_{con}}{\partial x_1} = K \cdot H(x_1 - x_2) \cdot \frac{3}{2}(x_1 - x_2)^{\frac{1}{2}}$$

$$\frac{\partial F_{con}}{\partial x_2} = -K \cdot H(x_1 - x_2) \cdot \frac{3}{2}(x_1 - x_2)^{\frac{1}{2}} \tag{14}$$

Since the vibroimpact system is periodically forced, changes of distances in the phase space direction $x_3 = \omega \cdot t$ are zero, what is manifested by the row of zeroes in the matrix J. Thus to find the largest Lyapunov exponent we use the projection of the phase space $(x_1, x_2, \dot{x}_1, \dot{x}_2, x_3)$ onto the phase plane $(x_1, x_2, \dot{x}_1, \dot{x}_2)$, that is the inner bracketed matrix in J (Eq. (13)) [12].

So far as step Heviside function $H(x_1 - x_2)$ is discontinuous one we must take into attention zero and nonzero for this function under integration both initial (Eq. (1)) and equations in variations (Eqs. (12) and (13)).

4.2   The Comparison of Different Estimations

We verified the estimation of the largest Lyapunov exponent by formulas (7), (8), (9), and by Benettin's algorithm for three different oscillatory regimes—periodic, quasiperiodic, and chaotic. Underline once more that it depends on initial conditions and may be found only after averaging of several results. We make conclusion about regime kind after the sign of Lyapunov exponent.

These numerical experiments showed the following.

- Difficulties under integration of equations in variations are the same ones as under integration of the initial equations because both one and the other contain the discontinuous Heviside function or its derivative. We avoid this difficulty by taking into attention zero and nonzero for it and using the program ode23s (MATLAB®ODE solvers) for integration the systems of stiff differential equations.
- Lyapunov exponents estimation by following the evolution of two nearby orbits in phase space allows to obtain their values roughly and to determine their signs.
- The calculations after formulas (7) and (8) are preferable. Formula (9) is described in some textbooks. None the less it gives worse results and we don't recommend its using.
- Of course, the calculation after Benettin's algorithm gives the better results. We have succeeded to estimate the largest Lyapunov exponent for strongly nonlinear nonsmooth discontinuous vibroimpact system after Benettin's algorithm and have

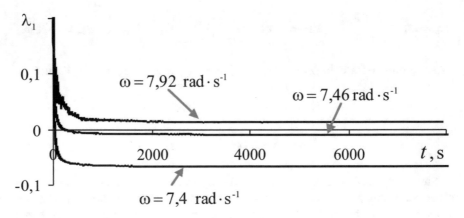

Fig. 11 The largest Lyapunov exponents for different oscillatory regimes

obtained sufficiently reliable results. The largest Lyapunov exponents calculated after Benettin's algorithm for different oscillatory regimes are shown at Fig. 11.

5 Quasiperiodic Route to Chaos

5.1 The Whole Motion Picture at Frequency Range $7.45\,\mathrm{rad}\cdot\mathrm{s}^{-1} \leq \omega \leq 8.0\,\mathrm{rad}\cdot\mathrm{s}^{-1}$

When we look at the whole motion picture at this frequency range we see the alternation of periodic regimes with long period and big impact number per cycle (which has got the name "chatter" or "rattle") with quasiperiodic and chaotic regimes. There are some transitional zones of prechaotic or postchaotic state. The appearance of subharmonic periodic vibrations is one characteristic precursor to chaotic motion. One of the signs of impending chaotic behaviour in dynamical system is a series of changes in motion nature as control parameter is varied.

It is very surprisingly, how so many different oscillatory regimes are arising in such small frequency range! How regime is changing qualitatively under the smallest frequency change (in the third significant figure, even in the fourth one)! Let us have a look at how it is happening, what vibroimpact system states are realizing in this narrow frequency range.

There are two plots which show the whole complicated motion picture very visibly and obviously (Figs. 12 and 13).

At Figs. 12 and 13 we see the change of system dynamic states when the control parameter is varied.

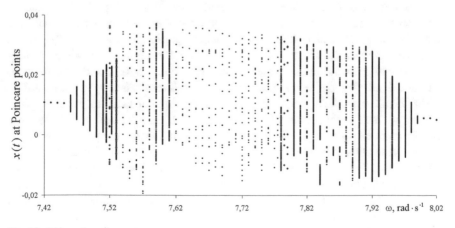

Fig. 12 Bifurcation diagram

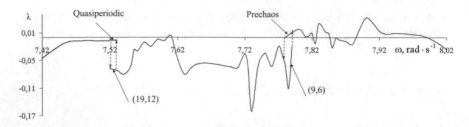

Fig. 13 The largest Lyapunov exponent dependence on control parameter

The bifurcation diagram is a widely used technique for investigation different states in a dynamical system as parameter is varied. At Fig. 12 the value of control parameter (a forcing frequency) is plotted on the horizontal axis and the values of phase coordinate $x_1(t)$ at Poincaré points are plotted on the vertical axis. There is only one value of one point coordinate in Poincaré map for (1,1)-regime, we see one point along vertical line at bifurcation diagram for $\omega < 7.46$ rad $\cdot$ s^{-1} and $\omega > 8.0$ rad $\cdot$ s^{-1}. There are n separate points along vertical line for (n,k)-periodic regimes. There are unbroken vertical lines for quasiperiodic and chaotic regimes.

At Fig. 13 the frequency ranges where the largest Lyapunov exponent is negative ($\lambda < 0$) correspond to periodic regimes, where $\lambda > 0$—to chaotic ones, and where $\lambda \approx 0$—to quasiperiodic oscillatory regimes.

5.2 Oscillatory Regimes at Frequency Range 7.45 rad $\cdot$ s$^{-1} \leq \omega \leq 8.0$ rad $\cdot$ s^{-1}

Let us now discuss more in details the route to chaos from quasiperiodic regime.

We observe the hysteresis effect (jump phenomenon) in very narrow frequency range 7.52 rad $\cdot$ s$^{-1} < \omega < 7.53$ rad $\cdot$ s^{-1}, where quasiperiodic and (19,12)-periodic regimes are coexisting. The arising of one or the other regime depends on history that is on initial conditions. It is the region where both periodic and quasiperiodic motions can coexist, and the precise motion that will result may be unpredictable.

The (19,12)-periodic regime is existing some more under short frequency varying.

Then we see the short zone of transition from (19,12)-periodic regime to other periodic regimes with long periods and big impact numbers per cycle ("chatter" or "rattle"). Transition is beginning at $\omega = 7.55$ rad $\cdot$ s^{-1} when one thirteenth impact is adding to 12 existing ones. Then there are regimes with very long periods, the Poincaré maps show big points numbers which after all form almost closed curve under 7.59 rad $\cdot$ s$^{-1} < \omega < 7.61$ rad $\cdot$ s^{-1}. The largest Lyapunov exponent is decreasing tenfold. So this regime looks highly quasiperiodic one if we look at its Poincaré map. But its Fourier spectrum (in logarithmic scale) is board and continuous, such as under chaotic motion (Fig. 14d). We think that we cannot call this regime both quasiperiodic and chaotic. It is transitional motion. At Fig. 14 we show phase

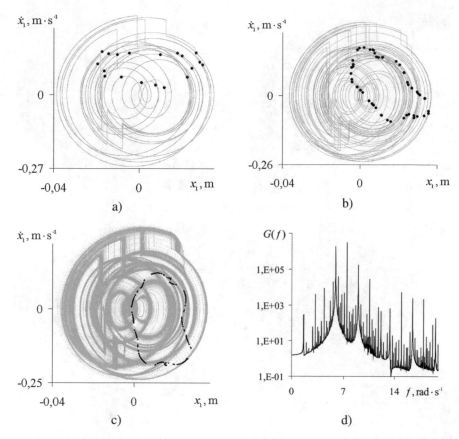

Fig. 14 Phase trajectories and Poincaré maps for: **a** (19,12)-periodic regime $\omega = 7.54$ rad · s^{-1}, $\lambda = -0.081$; **b** chatter $\omega = 7.58$ rad · s^{-1}, $\lambda = -0.020$; **c** transitional regime $\omega = 7.61$ rad · s^{-1}, $\lambda = 0.0027$; **d** Fourier spectrum for transitional regime $\lambda = 0.0027$

trajectories and Poincaré maps for (19,12)-periodic regime, chatter, and transitional regime.

When we are following the Figs. 12 and 13 we see the relatively big frequency range where periodic regimes are realising up to hysteresis phenomenon. There are subharmonics—(14,10) and (23,17)-periodic regimes which sharply replace each other under $\omega = 7.72$ rad · s^{-1}. Subharmonics play an important role in prechaotic vibrations so far as their appearance in frequency spectrum often is a characteristic precursor to chaotic motion. There may be in fact many patterns of prechaos behaviour. We again observe the hysteresis effect in very narrow frequency range 7.77 rad · s^{-1} < ω < 7.79 rad · s^{-1}, where transitional (prechaos) and (9,6)-periodic regimes are coexisting. In this frequency range the Poincaré map for transitional (prechaos) motion is becoming a set of points generally arranged in almost closed curve. It is the breakup of the quasiperiodic torus before the chaotic motion. At

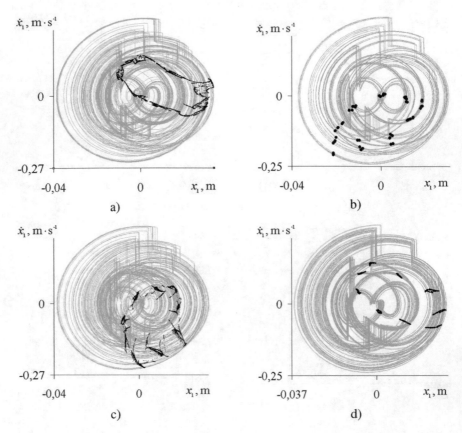

Fig. 15 Phase trajectories and Poincaré maps for: **a** $\omega = 7.80\,\text{rad} \cdot \text{s}^{-1}, \lambda = 0.018$; **b** $\omega = 7.82\,\text{rad} \cdot \text{s}^{-1}, \lambda = 0.0092$; **c** $\omega = 7.83\,\text{rad} \cdot \text{s}^{-1}, \lambda = 0.031$; **d** $\omega = 7.845\,\text{rad} \cdot \text{s}^{-1}, \lambda = 0.0086$

very narrow frequency range $7.80\,\text{rad} \cdot \text{s}^{-1} < \omega < 7.815\,\text{rad} \cdot \text{s}^{-1}$ there is a chaotic motion ($\lambda = 0.018$).

After that we see the zone with complicated motion picture under $7.80\,\text{rad} \cdot \text{s}^{-1} < \omega < 7.99\,\text{rad} \cdot \text{s}^{-1}$. Here chaotic motion alternates with prechaos and postchaos behaviour.

For example at Fig. 15 phase trajectories and Poincaré maps in such states are depicted.

Eventually there is really chaotic motion in narrow frequency range $7.90\,\text{rad} \cdot \text{s}^{-1} < \omega \leq 7.92\,\text{rad} \cdot \text{s}^{-1}$. At first we see how Poincaré map for quasiperiodic motion is deforming under $\omega = 7.93\,\text{rad} \cdot \text{s}^{-1}$ (Fig. 16) when frequency is decreasing. Chaotic motion is characterized by the breakup of the quasiperiodic torus structure as the control parameter is decreasing. When the torus is completely destroyed we observe the chaotic oscillations.

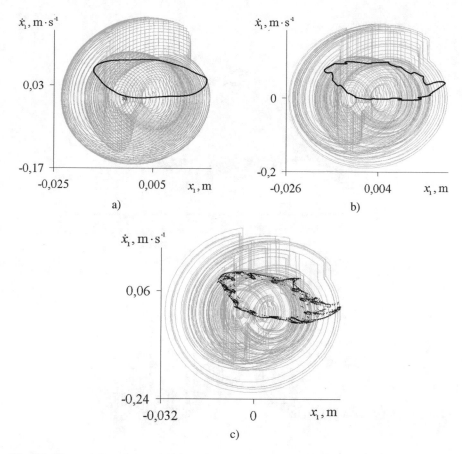

Fig. 16 Phase trajectories and Poincaré maps for: **a** $\omega = 7.94\,\mathrm{rad} \cdot \mathrm{s}^{-1}$, $\lambda = 0.0086$; **b** $\omega = 7.93\,\mathrm{rad} \cdot \mathrm{s}^{-1}$, $\lambda = 0.0069$; **c** $\omega = 7.92\,\mathrm{rad} \cdot \mathrm{s}^{-1}$, $\lambda = 0.014$

Let us have a look at chaotic motion under $\omega = 7.92\,\mathrm{rad} \cdot \mathrm{s}^{-1}$, $\lambda = 0.014$. At Fig. 17 Poincaré map for attached body m_2 and Fourier spectrum in logarithmic scale are depicted. Poincaré map does not consist of either a finite set of points or a closed orbit. Prof. F. Moon [12] have called his Poincaré map "Fleur de Poincaré", and we call our beautiful map as "Leaflet de Poincaré".

At Fig. 17 we see a broad continuous spectrum of Fourier components what is typical for chaotic motion. The generation of a continuous spectrum of frequencies is one of the characteristics of chaotic vibrations.

Our vibroimpact system is the damped one. Poincaré map is the singular characteristic of chaotic vibrations in such system. The Poincaré map appears as an infinite set of highly organized points arranged similar to parallel lines. Chaotic motion is not a formless chaos but one in which there is some order that is fractal structure.

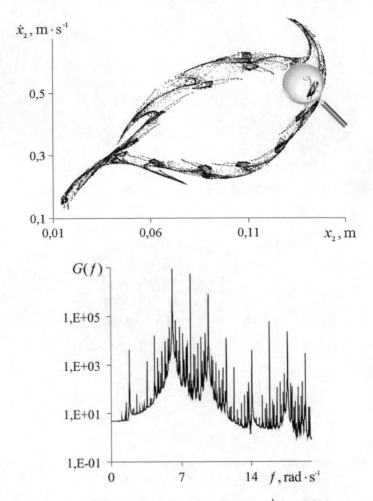

Fig. 17 Poincaré map and Fourier spectrum for $\omega = 7.92\,\mathrm{rad} \cdot \mathrm{s}^{-1}$, $\lambda = 0.014$

We enlarge a portion of the Poincaré map and observe further structure. We see that this structured set of points continues to exist after three enlargements (Fig. 18). So the motion appears to occur on the strange attractor. This embedding of structure within structure is a strong indicator of chaotic motion.

We observe the fractal structure of Poincaré map depicted at Fig. 17, at least this structure looks highly fractal. We have been able to obtain it when we have had 207000 Poincaré points on the map. It is shown at Fig. 18. We think that just this fractal structure implies the existence of a strange attractor.

Thus Poincaré map, Fourier spectrum, the largest positive Lyapunov exponent, and fractal structure of Poincaré map confirm the chaoticity of this regime.

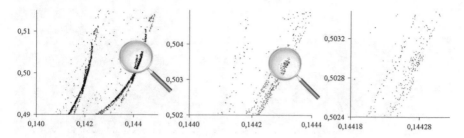

Fig. 18 Fractal structure

So we see three very short frequency ranges where chaotic motions are realizing: $7.80\,\text{rad} \cdot \text{s}^{-1} < \omega < 7.815\,\text{rad} \cdot \text{s}^{-1}, \omega = 7.83\,\text{rad} \cdot \text{s}^{-1}$, $7.90\,\text{rad} \cdot \text{s}^{-1} < \omega \leq 7.92\,\text{rad} \cdot \text{s}^{-1}$.

We think these chaotic motions may be considered as transient chaos because chaotic vibrations appear for some parameter changes and then almost abruptly degenerate into a quasiperiodic motion after a short time [12]. These chaotic sets don't attract trajectories from its vicinity, they are nonattracting ones. "Transient chaos is the form of chaos due to nonattracting chaotic sets in the phase space" [39, 40].

For comparison and for visualization the phase trajectories of three regimes are depicted in 3-th dimension picture at Fig. 19. These regimes are: (9,6)-periodic, quasiperiodic, and chaotic. There are the repeated identical layers at first plot. One can see the reversals of velocities under 6 impacts. The torus is seen at second plot, and disordered picture in limited space—at third one.

6   Conclusions

Dynamic behaviour of the nonlinear nonsmooth discontinuous vibroimpact system have been studied under changing of external force frequency and amplitude.

- Quasiperiodic route to chaos under changing of external force frequency was found as complicated one where oscillatory regimes have succeeded each other many times under very small control parameter varying in narrow frequency range. There were periodic subharmonic regimes—chatters, quasiperiodic, and chaotic regimes. There were the zones of transition from one regime to another, the zones of prechaotic or postchaotic motion. The hysteresis effects (jump phenomena) have been observed at two frequency ranges. There were three short frequency ranges where chaotic regime occurred. This chaos was found to be transient one because chaotic vibrations appeared for some parameter changes and then degenerated into a quasiperiodic motion after a short time.

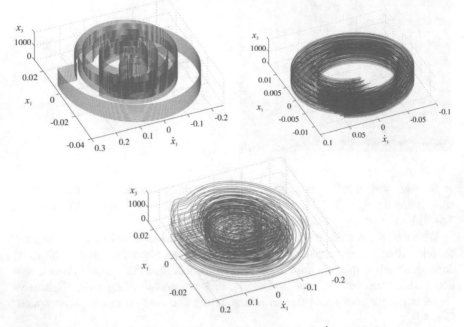

Fig. 19 Phase trajectories for (9,6)-periodic ($\omega = 7.78$ rad $\cdot$ s^{-1}, $\lambda = 0.062$), quasiperiodic ($\omega = 7.46$ rad $\cdot$ s^{-1}, $\lambda = -0.0091$), and chaotic ($\omega = 7.92$ rad $\cdot$ s^{-1}, $\lambda = 0.014$) regimes

- The chaoticity of obtained regime has been confirmed by typical views of Poincaré map and Fourier spectrum, by the positive value of the largest Lyapunov exponent, and by the fractal structure of Poincaré map.
- Discontinuous bifurcations occurred under changing of external force frequency and amplitude. At such points Floquet multiplier are leaving the unit circle by jump, sometimes by big jump. It is phenomenon unique for nonlinear nonsmooth dynamical system.
- So far as the calculation of Lyapunov exponents for discontinuous system has some difficulties we have tried to estimate the largest Lyapunov exponent by numerical integration two copies of dynamic system with nearby initial conditions and following the distance evolution between them. Three different formulas were used. Two of them allow to obtain the largest Lyapunov exponent values roughly and to determine their signs. But the third formula gives worse results and we don't recommend its using in spite of its description in some textbooks. The using of Benettin's algorithm provides the most reliable results.

References

1. Luo AC, Guo Y (2012) Vibro-impact dynamics. Wiley
2. Leine RI, Van Campen DH (2002) Discontinuous bifurcations of periodic solutions. Math Comput Model 36(3):259–273
3. Seydel R (2009) Practical bifurcation and stability analysis, vol 5. Springer Science & Business Media
4. Ivanov AP (2012) Analysis of discontinuous bifurcations in nonsmooth dynamical systems. Regul Chaotic Dyn 17(3–4):293–306
5. Leine RI, Van Campen DH, Van de Vrande BL (2000) Bifurcations in nonlinear discontinuous systems. Nonlinear Dyn 23(2):105–164
6. Di Bernardo M, Budd CJ, Champneys AR, Kowalczyk P, Nordmark AB, Tost GO, Piiroinen PT (2008) Bifurcations in nonsmooth dynamical systems. SIAM Rev 50(4):629–701
7. Brogliato B (ed) (2000) Impacts in mechanical systems: analysis and modelling, vol 551. Springer Science & Business Media
8. Alzate R (2008) Analysis and application of bifurcations in systems with impacts and chattering. Doctoral dissertation, Universita degli Studi di Napoli Federico II
9. Cheng J, Xu H (2006) Nonlinear dynamic characteristics of a vibro-impact system under harmonic excitation. J Mech Mater Struct 1(2):239–258
10. Volchenkov D, Leoncini X (eds) (2018) Regularity and stochasticity of nonlinear dynamical systems. Springer International Publishing
11. Kuznetsov SP (2001) Dynamical chaos. Fizmatlit, Moscow (2006), 356 pp
12. Moon FC (1987) Chaotic vibrations: an introduction for applied scientists and engineers. Research supported by NSF, USAF, US Navy, US Army, and IBM. Wiley-Interscience, New York, 322 pp
13. Luo AC (2016) Periodic flows to chaos in time-delay systems, vol 16. Springer
14. Afraimovich V, Machado JAT, Zhang J (eds) (2016) Complex motions and chaos in nonlinear systems. Springer International Publishing
15. Shvets AY, Sirenko VA (2015) New ways of transition to deterministic chaos in nonideal oscillating systems. Res Bull Natl Tech Univ Ukr "Kyiv Polytech Inst" 1(99):45–51
16. Serweta W, Okolewski A, Blazejczyk-Okolewska B, Czolczynski K, Kapitaniak T (2014) Lyapunov exponents of impact oscillators with Hertz's and Newton's contact models. Int J Mech Sci 89:194–206
17. Manneville P, Pomeau Y (1980) Different ways to turbulence in dissipative dynamical systems. Phys D: Nonlinear Phenomena 1(2):219–226
18. Benettin G, Galgani L, Giorgilli A, Strelcyn JM (1980) Lyapunov characteristic exponents for smooth dynamical systems and for Hamiltonian systems; a method for computing all of them. Part 1: theory. Meccanica 15(1):9–20
19. Muller PC (1995) Calculation of Lyapunov exponents for dynamic systems with discontinuities. Chaos Solitons Fractals 5(9):1671–1681
20. Stefanski A, Dabrowski A, Kapitaniak T (2005) Evaluation of the largest Lyapunov exponent in dynamical systems with time delay. Chaos Solitons Fractals 23(5):1651–1659
21. De Souza SL, Caldas IL (2004) Calculation of Lyapunov exponents in systems with impacts. Chaos Solitons Fractals 19(3):569–579
22. Ageno A, Sinopoli A (2005) Lyapunov's exponents for nonsmooth dynamics with impacts: stability analysis of the rocking block. Int J Bifurc Chaos 15(06):2015–2039
23. Andreaus U, Placidi L, Rega G (2010) Numerical simulation of the soft contact dynamics of an impacting bilinear oscillator. Commun Nonlinear Sci Numer Simul 15(9):2603–2616
24. Komuro M, Kamiyama K, Endo T, Aihara K (2016) Quasi-periodic bifurcations of higher-dimensional tori. Int J Bifurc Chaos 26(07):1630016
25. Afraimovich VS, Shilnikov LP (1991) Invariant two-dimensional tori, their breakdown and stochasticity. Am Math Soc Transl 149(2):201–212
26. Shilnikov A, Shilnikov L, Turaev D (2004) On some mathematical topics in classical synchronization: a tutorial. Int J Bifurc Chaos 14(07):2143–2160

27. Bakri T (2005) Torus breakdown and chaos in a system of coupled oscillators. Int J Non-Linear Mech
28. Verhulst F (2016) Torus break-down and bifurcations in coupled oscillators. Procedia IUTAM 19:5–10
29. Bazhenov VA, Lizunov PP, Pogorelova OS, Postnikova TG, Otrashevskaia VV (2015) Stability and bifurcations analysis for 2-DOF vibroimpact system by parameter continuation method. Part I: loading curve. J Appl Nonlinear Dyn 4(4):357–370
30. Bazhenov VA, Lizunov PP, Pogorelova OS, Postnikova TG (2016) Numerical bifurcation analysis of discontinuous 2-DOF vibroimpact system. Part 2: Frequency-amplitude response. J Appl Nonlinear Dyn 5(3):269–281
31. Bazhenov VA, Pogorelova OS, Postnikova TG (2017) Stability and discontinious bifurcations in vibroimpact system: numerical investigations. LAP LAMBERT Academic Publishing GmbH and Co., KG Dudweiler, Germany
32. Lamarque CH, Janin O (2000) Modal analysis of mechanical systems with impact non-linearities: limitations to a modal superposition. J Sound Vib 235(4):567–609
33. Bazhenov VA, Pogorelova OS, Postnikova TG (2016) Dangerous bifurcations in 2-DOF vibroimpact system. Bull Natl Tech Univ "Kharkiv Polytech Inst" (26):109–113
34. Zakrzhevsky M, Schukin I, Yevstignejev V (2007) Rare attractors in driven nonlinear systems with several degrees of freedom. Transp Eng 24
35. Lichtenberg AJ, Lieberman MA (2013) Regular and stochastic motion, vol 38. Springer Science & Business Media
36. Chirikov BV, Shepelyanskii DL (1982) Dynamics of some homogeneous models of classical Yang-Mills fields. Sov J Nucl Phys 36(6):908–915
37. Wolf A, Swift JB, Swinney HL, Vastano JA (1985) Determining Lyapunov exponents from a time series. Phys D: Nonlinear Phenom 16(3):285–317
38. Bazhenov VA, Pogorelova OS, Postnikova TG (2018) Lyapunov exponents estimation for strongly nonlinear nonsmooth discontinuous vibroimpact system. In: Strength of materials and theory of structures (99) (in press)
39. Lai YC, Tél T (2011) Transient chaos: complex dynamics on finite time scales, vol 173. Springer Science & Business Media
40. Afraimovich VS, Neiman AB (2017) Weak transient chaos. Advances in dynamics, patterns, cognition. Springer, Cham, pp 3–12

Modeling Ensembles of Nonlinear Dynamic Systems in Ultrawideband Active Wireless Direct Chaotic Networks

A. S. Dmitriev, R. Yu. Yemelyanov, M. Yu. Gerasimov, and Yu. V. Andreyev

Abstract Active wireless networks are considered in this paper as an instrument for modeling various ensembles of coupled nonlinear systems. Active wireless network is a generalized multi-element wireless processor platform, whose elements interact via radio channels. The nodes of active network comprise a transceiver and an actuator. Actuator's microprocessor solves equations of the simulated dynamic system, while wireless direct chaotic transceivers enable couplings between dynamic systems of the ensemble.

1 Introduction

A classical *wireless sensor network (WSN)* is a set of nodes with sensors, combined in ensemble by means of radio channel connections [20]. However, there are more and more examples of wireless networks whose nodes (besides a transceiver and a sensor, or several sensors) comprise actuators, i.e., devices that act on the environment, visualize data (e.g., with LED or LCD), and/or process information (e.g., with a microcontroller unit (MCU) or a processor) [2]. Below, such generalized wireless sensor networks will be called *active wireless networks (AWN)*.

Wide range of equipment in the nodes along with communication capability allows us to consider AWN not only as a tool for acquiring and delivering data from a certain area to a processing center, but as a powerful technological platform for solving tasks related with multi-element interacting systems.

In this paper we show that AWNs can be used to model interacting dynamic systems.

A. S. Dmitriev (✉) · M. Yu. Gerasimov · Yu. V. Andreyev
V.A. Kotelnikov Institute of Radio Engineering and Electronics of the RAS, Mokhovaya 11-7, Moscow 125009, Russia
e-mail: chaos@cplire.ru

R. Yu. Yemelyanov · M. Yu. Gerasimov · Yu. V. Andreyev
Institute of Physics and Technology (State University), 9 Institutskiy per., Dolgoprudny 141700, Russia

© Higher Education Press 2021   47
D. Volchenkov (ed.), *Nonlinear Dynamics, Chaos, and Complexity*,
Nonlinear Physical Science, https://doi.org/10.1007/978-981-15-9034-4_4

When AWN is used as a modeling tool, element interconnections via radio channels provide any necessary network topology. This feature, important for active network to be flexible as a modeling instrument, is achieved by means of proper addressing of the transmitted signals.

2 Historical Background

One of the research directions close to the theme of this paper is Cellular Neural Networks (CNN) [3, 4], which are based on two basic concepts: local connections of nonlinear elements and analog circuit dynamics. In CNN, one can use various nonlinear elements: static, described with nonlinear functions, as well as dynamic, e.g., described with differential equations. CNN is assumed to comply homogeneity condition, i.e., all its nodes are assumed equal. The theory and applications of cellular neural networks were developing intensively, gradually involving still wider range of problems. Then cellular neural networks were generalized to cellular *nonlinear* networks (also abbreviated as CNN) [15]. The task was formulated as the design of an algorithmically programmable analog real-time one-chip cellular computer, possessing the power of a supercomputer. Unlike cellular neural network, cellular nonlinear network was programmable, it had global distributed analog memory and logics, and the network elements were supposed to exchange data at a high rate, e.g., using electromagnetic waves. A remarkable set of potential CNN applications was also described in [15], including neural-like calculations, "programmable physics", "programmable chemistry", and "programmable biology".

Relations of WSN and CNN were analyzed in [10]. As was noted there, in large-scale wireless networks due to energy and interference limitations the communication reach of a node grasps only its nearest neighbors, and the control and resource interchange algorithms are stored locally in each node in distributed form. With these features, a wireless system of locally connected nodes can be certainly considered as a CNN, which allows us to use copious CNN results by analysis and design of WSN and its generalizations.

Haenggi also concluded that ubiquitous wireless microsensor and actuator tools can essentially extend our understanding and control of complex physical systems [10]. Possibility of overall physical monitoring and control creates an all-new situation in almost all scientific disciplines. Networks of such devices can constitute a new computation platform with amazing capabilities.

Active wireless networks are a generalization of WSN, so, they inherit all the features considered in relation analysis of WSN and CNN. However, they have a number of new properties. Let us see what is common and what is different in CNN and AWN.

The common features:

1. Both networks are multi-element structures;
2. Each element has an information processing block;

3. The elements are combined with connections through which they interact; these connections are often space-limited and local.

Essential differences:

1. The connection structure of CNN as a universal machine and supercomputer inherits topological structure of cellular neural (nonlinear) networks and mostly contains local spatially uniform links, described with a certain template. This restriction was specially imposed in [3, 4, 15], so that CNN could be implemented as one chip, which was practically impossible for general neural-like systems, in which the number of connections was excessively high.
2. CNN is a mostly computational structure, its definition includes no sensor or actuator functions. Of course, this doesn't exclude a possibility to add such functions, but there are natural restrictions on this way: one-chip implementation of a system of sensors as it is understood in sensor networks, has no sense in CNN, because numerous sensors would be located in the same space position. Otherwise, if CNN cells are dispersed over a large territory, then the network is no longer a one-chip universal machine and supercomputer.
3. From the viewpoint of modeling, CNN can solve various tasks and problems with only local connections, with no interaction with the environment.

Active wireless networks can be used to solve special sensor problems, to model and simulate many scientific and technical problems, including those that can't be solved with CNNs, i.e., AWNs can implement:

1. Networks with arbitrary connection topology;
2. Networks with heterogeneous nodes;
3. Networks that interchange data with the environment;
4. Networks that act on the environment.

3   Problem Statement and Modeling Scheme

Let there be an ensemble of interacting dynamic systems, each described with evolution equations, e.g., a system of nonlinear differential equations. The task is to realize the ensemble on AWN and to use this realization to observe and investigate the ensemble dynamics.

The first step of ensemble realization on AWN is representation of the interacting dynamic systems. This representation can be divided into several stages (Fig. 1):

1. Programming equations of each ensemble element on the processor of the AWN node corresponding to this element;
2. Determining and establishing connections between active network nodes, corresponding to couplings between the ensemble elements;
3. Realization of the connections with radio channels.

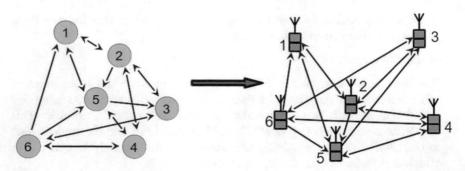

Fig. 1 The structure of a fully-connected ensemble of dynamic systems and its mapping to a wireless network

Programming ensemble equations on the processor is done in three steps: writing and debugging a high-level language program, program compilation to machine code, loading the program code to processor memory.

A key feature of AWN as a model of interacting dynamic systems is implementation of node connections with radio channels, either directly or through relays. This means that radio channels allow us to simulate ensembles with arbitrary connection topology. Note that the topology can easily be changed, because it is defined by node programming.

Consider an active network with all the nodes within direct radio coverage. This means that each node can exchange data with the nodes with which it must be connected. In this case, to transmit data from node i to other nodes, i-th node broadcasts a packet with information about itself. All nodes receive this packet, and the nodes that are coupled in the mathematical model with the ensemble element represented by i-th node, use the packet data to form their own states on the next time step. The nodes that are not coupled in model with the transmitting node simply discard the received packet.

In each node, the processor integrates equations that describe the behavior of the corresponding ensemble element. All nodes operate independently, making each integration step on time interval Δt. At the beginning of interval Δt, each network element transmits a message about the state of its variables and then switches to receiver mode till the end of this interval. As a result, each node during its "own" time interval Δt transmits its state data to all the nodes with which it is coupled and receives data on the states of those nodes. To visualize the state of the node variable, its value is mapped to LED color of the actuator node.

Though the nodes do not transmit signals continuously (one packet in Δt interval), at each integration step they exchange the state data with all the other nodes at each integration step, so from the modeling viewpoint this actually corresponds to continuous coupling.

4   Ultrawideband Direct Chaotic Active Network

Consider a realization of AWN on ultrawideband (UWB) direct chaotic transceivers [5, 7, 8]. The choice of these transceivers is explained by better multipath propagation characteristics (hence, higher reliability).

Up to now, several types of UWB direct chaotic transceivers were developed. Here, a UWB direct chaotic transceiver TR-43 is used in active network nodes.

The transceiver layout is presented in Fig. 2a. Microwave part of the transmitter is implemented with a transistor chaotic generator. To form chaotic radio pulses, generator signal is internally modulated [7, 8]. Omnidirectional transceiver antenna is fabricated on the same device PCB (Fig. 3a). Envelope receiver is implemented with log detector (operation range 1 MHz to 10 GHz).

The transceiver performance is given in Table 1.

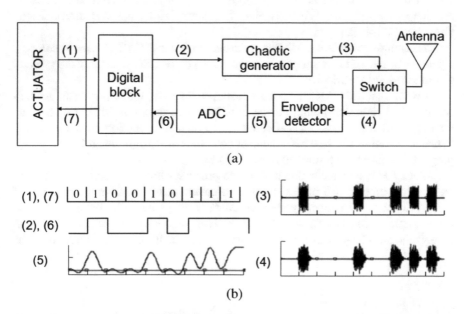

(a)

(b)

Fig. 2 a Layout of ultra-wideband direct chaotic transceiver with connected actuator, (ADC—analog-to-digital convertor); **b** Signal forms

Table 1   UWB transceiver properties

Output signal band	3.0–5.0 GHz
Average emission power at 3 Mbps	0 dBm (–33 dBm/MHz)
Average emission power at 0.1 Mbps	–21 dBm (–48 dBm/MHz)
Distance	25–30 m
Maximum data rate (PHY)	6 Mbps
Voltage supply	4.5 VDC

4.1 Direct Chaotic Transceiver

The nodes of the active network communicate with each other using wireless UWB transceivers. Transmitted information is encoded by chaotic radio pulses using chaotic on-off keying (chaotic OOK, COOK): chaotic radio pulse on a prescribed time position, if bit "one" is transmitted, or no pulse in case of "zero" bit (Fig. 2b). To compensate effects of multipath propagation, a small void guard interval is left between the pulses.

4.2 Network Node

Each node of the active network comprises a wireless transceiver and a special actuator board (Fig. 3). The transceiver is represented by a UWB direct chaotic transceiver TR-43 [5].

The actuator comprises a STM32L microcontroller unit (MCU), used to compute equations of one of the ensemble elements, and a color LED to display the value of one of the element variables.

STM32L is a RISC MCU, clock frequency 1–32 MHz, rather powerful, yet very compact and little consuming device. What is important, this MCU can emulate floating-point operations, thus, it can be used as a full-function device to model dynamic systems. The ensemble element equations are programmed in C, then the program is converted (compiled) to MCU code.

Color LED is a visualization unit on the processor board, its color directly represents the value of one of the ensemble element variables. Simultaneous use of several LEDS in the network gives a visual dynamic picture of collective behavior. For example, such a color picture helps clearly distinguish such phenomena as synchronization or clustering of ensemble elements. Actually, color LED display of the

(a)   (b)

Fig. 3 Ultrawideband direct chaotic transceiver TR-43 and actuator with RGB LED

variable value is a hardware implementation of the method of state-to-color mapping widely used in studies of dynamic systems.

UWB transceiver and actuator are connected with UART (universal asynchronous receiver-transmitter) interface (data rate $C = 46,080$ bps).

4.3   Organization and Operation of UWB Network

The simulated dynamic systems interact by means of sending data packets between the active network nodes. The structure of the network packets is shown in Fig. 4. The packet contains the sender ID field (4 bytes), the variable value (values) of the simulated dynamic system (data field) and the checksum. Unique IDs are used to identify network devices, to determine the packet sender and to virtually form the necessary network topology. The value of the simulated dynamic system variable is used to compute its equations. Total packet length in the examples below is 16 bytes.

Consider data exchange in active network. Let there be n nodes that are turned on at arbitrary moments. After turn-on, each node is in the receiver mode for time T_1, then it integrates equations of the simulated dynamic system (Euler method) for time T_2, and for time T_3 it transmits data on the dynamic system state; then the cycle is repeated (the cycle duration is $T = T_1 + T_2 + T_3$).

All nodes are assumed to be within radio coverage of each other; time intervals $T_2, T_3 \ll T$; turn-on moments can be considered random. Packet collision probability P is proportional to $T_3/T \ll 1$, which allows us to neglect packet collisions.

In $2T$ after the last node turn-on, each node will have data on the states of all the other network nodes, and this information will be updated on each its interval T.

5   Modeling Continuous-Time Systems

Let there be an ensemble of interacting dynamic systems, each described with one or several nonlinear differential equations. The task is to realize the ensemble on AWN and to use this realization to observe and investigate the ensemble dynamics. As a model example of a continuous-time system, let us take Kuramoto ensemble [11, 12].

Fig. 4 Layout of network packets for data exchange on AWN

Sender ID (4 Bytes)	State value data (8 Bytes)	Checksum (4 Bytes)

5.1 Kuramoto Model

Kuramoto model is a system of N coupled phase oscillators with natural frequencies ω_i, distributed with a given probability density $g(\omega)$, and described with equations:

$$\dot{\theta}_i = \omega_i + \sum_{j=1}^{N} K_{ij} \sin(\theta_j - \theta_i), \ i = 1, \ldots N. \tag{1}$$

Each oscillator tries to oscillate independently on its natural frequency, while the couplings tend to synchronize it with the others. If the couplings are sufficiently weak, oscillators operate incoherently; if a certain threshold is exceeded, the nodes spontaneously synchronize.

First [11, 12], synchronization analysis of Kuramoto system was accomplished for the case of mean-field couplings, i.e., in assumption that $K_{ij} = K/N > 0$ in Eq. (1). The model equations were rewritten in a more convenient form with an order parameter

$$re^{i\psi} = \frac{1}{N} \sum_{j=1}^{N} e^{i\theta_j}, \tag{2}$$

where $0 \le r(t) \le 1$ is a coherence degree of the oscillator set, and $\psi(t)$ is the average phase. Using this parameter, Eq. (1) takes the form

$$\dot{\theta}_i = \omega_i + Kr \sin(\psi - \theta_i), \ i = 1, \ldots N. \tag{3}$$

Evidently, each oscillator is coupled with the common average phase $\psi(t)$ with coupling strength Kr.

In the mean-field form (3), the model character becomes obvious. Each oscillator is uncoupled with all the others, however, they interact through the mean field, which is determined by r and ψ. In particular, phase θ_i is pulled to the average phase, not to the phase of a concrete oscillator. Moreover, effective interaction strength is proportional to coherence r. This proportionality establishes positive feedback between interaction strength and coherence: as the ensemble becomes more coherent, r grows, thus coupling efficiency Kr increases, which draws, as a rule, more oscillators in the synchronized cluster. If coordination increases due to new elements drawn, the process will continue; otherwise, it limits itself. This mechanism, the basis for spontaneous synchronization, has been discovered by Winfree [21], and it exhibits itself especially bright in Kuramoto model.

Kuramoto model was introduced and investigated in the case of a large number of phase oscillators, however, it proved interesting also in the case of small ensembles.

First of all, here we talk about discovery and investigation of dynamic chaos in phase oscillator ensembles [13, 14], and about investigation of the states of chimera type [1].

For instance, as was shown in [13, 14], with an increase of coupling parameter K, chaotic behavior of Kuramoto model is observed even in the case of $N = 4$ elements. Chaos is preceded by quasiperiodic behavior. With an increase of N, hyperchaos can take place: even in the cases of $N = 6$ and 7 there are modes with two positive Lyapunov exponents. The authors investigated Kuramoto ensembles with higher number of elements and found that the number of positive Lyapunov exponents can achieve half the number of oscillators. Yet the strongest phase chaos is observed in medium-size ensembles.

Thus, Kuramoto model demonstrates various sophisticated dynamic phenomena, both in the case of small and large number of elements. It is a good qualitative model for a wide range of natural phenomena, such as synchronization of firefly swarms [9], Josephson effect [19], coupled chemical reactions [18], neural population physiology [17], etc. One can also note the use of this model to explain oscillations of Millennium bridge in London during the opening ceremony [16].

From the viewpoint of modeling on active wireless networks, another argument in favor of Kuramoto model is that each its element is described by only one first-order differential equation.

5.2 Emulation of Kuramoto Model on the Active Wireless Network

Emulation of collective behavior of phase oscillators of Kuramoto model on active wireless network includes the following stages: network initiation, transient process and stationary dynamic mode.

The *network initiation* is accomplished by means of turning all the nodes on.

Transient process. Network operation begins with a transient process. Each AWN node comprises a wireless transceiver and an actuator, that were described above. Dynamic system of ith node is defined by Eq. (4). Actuator of ith node begins integration of Eq. (4) with parameter values corresponding to the investigated model. All nodes have their own initial values and all start at different moments. That is why in the beginning a transient process is observed, during which the dynamic systems converge from arbitrary initial conditions to a stationary state. The value of phase oscillator variable θ_i, that oscillates in the range $(0, 2\pi)$, is depicted by RGB LED of ith node actuator. The nodes, each on its own time interval T, communicate with each other by radio and exchange data about their states. The obtained variable values are used in integration with weights K/N.

Stationary dynamic mode. After a certain number of integration steps the system converges to a stationary dynamic mode.

5.3 Computer Simulation

Before experimenting with Kuramoto ensemble using the active wireless network, an ensemble of N elements was simulated on PC and dynamics of the ensemble was analyzed. The physical experiment was later carried out with $N = 6$ devices, so, the ensemble with the same number of elements was simulated. The ensemble equations were

$$\dot{\theta}_i = \omega_i + \frac{K}{6} \sum_{j=1}^{N} \sin(\theta_j - \theta_i), \; i = 1, \ldots 6. \tag{4}$$

Ensemble dynamics was analyzed as a function of coupling strength K; the values of natural frequencies were set at:

$$\omega_i = -1 + \frac{2}{5}(i - 1), \; i = 1, \ldots 6 \tag{5}$$

where i was the oscillator number. Oscillation mode diagram and the largest Lyapunov exponent vs coupling strength were plotted (Fig. 5).

Three different types of dynamic modes can be distinguished in the ensemble behavior: quasiperiodic oscillations (for weak coupling), dynamic chaos, and synchronization. Besides, in the absence of couplings the mode of independent oscillations is observed.

In case of symmetric distribution of natural frequencies (5), the synchronous mode degenerates into equilibrium state with zero oscillation frequency. Based on the simulation results, for experiments we have taken the following values of parameter K: $K = 0$ for the mode of independent oscillations; $K = 1.3$ for dynamic chaos, and $K = 3$ for synchronous mode.

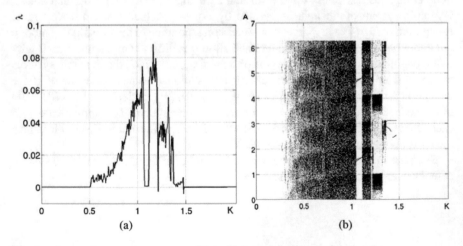

(a)   (b)

Fig. 5 **a** Largest Lyapunov exponent and **b** oscillation mode diagram for $N = 6$

6 Experiments with Network

In the experiments, $N = 6$ nodes were placed on the table. The devices were within direct radio coverage and were turned on one by one with several seconds intervals.

The dynamic system parameters of each AWN node were set at the stage of compiling the actuator program. This program provided: computation of phase oscillator dynamic system by means of periodic integration of the oscillator equations with regard to the phases of the other oscillators of the ensemble; transceiver control, i.e., switching to transmitter/receiver modes; and control of the LED color according to the oscillator phase. The idea of mapping oscillator phase values to LED colors is illustrated in Fig. 6.

Initial phase of each oscillator was set random. Natural frequencies of the oscillators were set the same as in the computer simulation (see Eq. (5)).

For convenient observation, Eq. (4) were integrated once in $T = 100$ ms (period of integration cycle). In this case, the variable value and the corresponding LED color changed evidently on several seconds time interval.

As is known, wireless communications is less reliable than wire lines, and sometimes data packets can be lost, e.g., due to bad propagation conditions, packet collisions, etc. In our case, this means that a node can receive less information than necessary, i.e., data from some other nodes is lost at a certain integration step. For this situation, we provided the following correction mechanism. If the node receives data on the phases of M ($M < N - 1$) other dynamic systems during the integration cycle, then this particular node on this integration cycle solves Kuramoto problem for the ensemble of ($N = M + 1$) elements and only the received phase values are taken into account. In particular, when there is no information from the other nodes, the term in Eq. (4) that corresponds to the sum of couplings between the nodes becomes zero.

So, a packet loss can cause certain integration fault, however, our experiments show that the rate of such events is negligibly small.

Fig. 6 Mapping phase values to colors

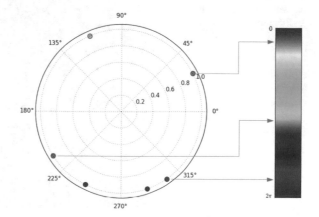

6.1   System Dynamics in the Absence of Couplings

First experiment, $K = 0$ (Fig. 7). The nodes are constantly exchanging phase values, however, the term with the sum in Eq. (4), corresponding to couplings between dynamic systems, is equal to 0. Therefore, all dynamic systems on the nodes operate independently, with different natural frequencies. The LED color of each AWN node is changing gradually and strictly periodically; no correlation of the LEDs' colors is observed.

6.2   Dynamic Chaotic Mode

Coupling strength $K = 1.3$ (Fig. 8) corresponds to dynamic chaos in computer simulation of Kuramoto model. In this mode, Lyapunov exponent is positive and equal approx. to 0.03. As compared with the previous experiment, the ensemble behavior is visually different, LEDs' colors change aperiodically, irregularly, with no visible correlation of different nodes.

6.3   Equilibrium State

In the third experiment, coupling strength is $K = 3$. After turn-on, the devices begin operating at random colors, then the process converges to an equilibrium state, in which all the LEDs constantly emit lights of the same color.

Fig. 7 Emulation of Kuramoto ensemble on active wireless network—independent oscillations

Fig. 8 Emulation of
Kuramoto ensemble on
active wireless
network—dynamic chaotic
mode

6.4 Synchronization

In the fourth experiment (Fig. 9) $K = 3$. Natural oscillator frequencies are set random
in the range $(-1, 1)$. In this case, synchronization mode with nonzero resulting
frequency is observed. Visually, the process looks like a coherent periodic color
variation. The frequency of coherent oscillations is different in each new turn-on.

Fig. 9 Emulation of
Kuramoto ensemble on
active wireless
network—synchronization
mode

6.5 Destruction of the Couplings

In the last experiment, we used AWN with the same parameters as in the fourth experiment. When the ensemble synchronized, near the ensemble we turned on a special direct chaotic transmitter (Fig. 10), which emitted a strong continuous chaotic signal in the frequency band of AWN transceivers. This radio interference destroyed all AWN couplings and stopped data exchange in the ensemble. As a result, LEDs' colors became different and started changing independently, i.e., the ensemble desynchronized. When the transmitter of interference signal was turned off, synchronization quickly restored.

In Fig. 11 the dynamics of the phases of individual oscillators during this experiment are presented (for better clarity, phase derivatives are shown). The phase values were obtained by means of "eavesdropping the conversation" of the ensemble elements. Actually, a supplementary UWB direct chaotic receiver intercepted data packets transmitted by the ensemble transceivers, extracted the phase values from the packets, and transferred phase data to personal computer, where it was analyzed.

In Fig. 11, the moments of switching the external interference on and off are shown. All three stages (synchronization, desynchronization, and synchronization again) can easily be seen in this figure.

7 Limits of Modeling on Active Wireless Networks

The above experimental AWN consisted of six nodes. The question is: how large size an AWN can have in such experiments, what is the maximum number N of its nodes?

This number is determined by the following factors: the rate of data transmission through radio channel between transceivers, the rate of data transmission through

Fig. 10 Emulation of Kuramoto ensemble on active wireless network—destruction of coupling using external interference

Fig. 11 Phase dynamics of
Kuramoto ensemble in the
experiment with coupling
destruction

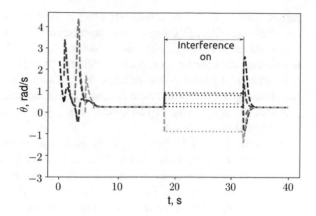

'transceiver–actuator' interface (Fig. 2), probability of packet collisions in communication channels, and the rate of dynamic system computation. Let us analyze the effects of these factors.

Rate of data transmission through radio channel. To exchange data with all AWN nodes during integration cycle T, each node must transmit one data packet with its state and receive $N - 1$ packets from its neighbors, the packet length being L. The total amount of transmitted and received information I that a channel with capacity (throughput) C can pass through over time T is

$$I = CT.$$

In the limit case, this amount of information is equal to information in N data packets sent by the maximum possible number of devices on the network, i.e.,

$$I = NL = CT. \tag{6}$$

Hence, the maximum possible number of the network nodes with regard to communication channel throughput is equal to

$$N = \frac{CT}{L}. \tag{7}$$

Let us make some evaluations. The data rate of TR-43 transceivers is $R = 6$ Mbps. So, as follows from Eq. (7), for the packet length $L = 128$ bit (16 bytes) and cycle period $T = 100$ ms, the maximum number of AWN nodes is above 4000.

Data rate of "transceiver–actuator" interface. All packets transmitted and received by the transceiver are transferred through this interface. Hence, Eq. (7) can be used to evaluate the maximum admissible number of AWN devices regarding the rate of communication interface. In the experimental testbeds, this interface is represented by UART with data rate $C = 46{,}080$ bps. According to Eq. (7), due to low data rate of UART the maximum number of AWN nodes is no more than 30.

Probability of packet collisions in communication channels. In the case of 30 (at maximum) network devices, packet collision probability is negligibly small. However, even if a packet is lost because of collisions or other factors, then due to a small change of oscillator phase on one integration step, the phase values from the previous integration step can be used in Eq. (4), with practically no degradation of the computation accuracy.

Computation rate by modeling a dynamic system on MCU. The time necessary to compute one integration step of Kuramoto model with six elements is 3–4 μs. For Eq. (4) the amount of calculations in each actuator increases linearly with the number of network nodes. Since for six nodes the integration time is no more than 3–4 μs, adding each new node adds no more than 1 μs to integration time. Assuming that computation time is less than 10% of period $T = 100$ ms, we obtain the limit number of nodes (regarding only this factor) as approx. 10,000.

Thus, the main factor of the described hardware platform that limits the number of AWN nodes is the rate of data transmission through "transceiver–actuator" interface. If slow UART is replaced by high-rate SPI (Serial Peripheral Interface), the maximum number node is increased many times.

8   Conclusions

In this paper, we investigated the idea of implementing ensembles of coupled nonlinear discrete- or continuous-time dynamic systems with active wireless networks, i.e., ensembles of calculating machines connected together with wireless radio channels.

For this purpose, an UWB wireless communication network was equipped with specially designed actuators, capable of integrating system equations and exhibiting system state dynamics with LED color variations. MCU software was developed that provided actuator operation as a part of wireless active node and AWN operation as a multi-element continuous-time dynamic system. Such an AWN can be treated as an implementation of a *cellular nonlinear network* [15].

To demonstrate capabilities of AWN as a universal platform for modeling ensembles of coupled nonlinear systems, a Kuramoto ensemble was taken for experiments. On example of this dynamic system with six elements we showed that the dynamics of active wireless network could fully conform with the dynamics of the original mathematical model.

The obtained results prove that AWN can effectively be used for experimental emulation of multi-element continuous-time (and discrete-time [6]) dynamic systems.

Acknowledgements This research was financed by a grant of the Russian Science Foundation (Project No. 16-19-00084).

References

1. Abrams D, Strogatz S (2004) Chimera states for coupled oscillators. Phys Rev Lett 93(17):174102
2. Chen J, Johansson KH, Olariu S, Paschalidis IC, Stojmenovic I (2011) Special issue on wireless sensor and actuator networks. IEEE Trans Autom Control 56(10)
3. Chua L, Yang L (1988) Cellular neural networks: applications. IEEE Trans Circuits Syst 35(10):1273–1290
4. Chua L, Yang L (1988) Cellular neural networks: theory. IEEE Trans Circuits Syst 35(10):1257–1272
5. Dmitriev A, Efremova E, Lazarev V, Gerasimov M (2013) Ultra wideband wireless self-organizing direct chaotic sensor network. Achiev Mod Radioelectron 3:19–29
6. Dmitriev A, Gerasimov M, Yemelyanov R, Itskov V (2015) Ensembles of dynamic systems in active wireless networks. J Commun Technol Electron 60(1):69–74
7. Dmitriev A, Kyarginsky B, Maximov N, Panas A, Starkov S (2000) Perspectives of direct chaotic communication systems creation in radio and microwave ranges. Radioengineering 3:9–20
8. Dmitriev A, Kyarginsky B, Panas A, Starkov S (2001) Direct chaotic communication systems in microwave range. Radio Eng Electron 46(2):224–233
9. Ermentrout GB (1991) An adaptive model for synchrony in the firefly pteroptyx malaccae. J Math Biol 29:571–585
10. Haenggi M (2003) Distributed sensor networks: a cellular nonlinear network perspective. Int J Neural Syst 13(06):405–414
11. Kuramoto Y (1975) Self-entrainment of a population of coupled non-linear oscillators. In: Araki H (ed) International symposium on mathematical problems in theoretical physics, vol 39. Springer, Berlin, pp 420–422
12. Kuramoto Y (1984) Chemical oscillations, waves and turbulence. Springer, Berlin
13. Maistrenko Y, Popovych O, Tass P (2005) Desynchronization and chaos in the Kuramoto model. Lecture notes in physics, vol 671, pp 285–306
14. Popovych OV, Maistrenko YL, Tass PA (2005) Phase chaos in coupled oscillators. Phys Rev E 71(6):065201
15. Roska T, Chua LO (1993) The CNN universal machine: an analogic array computer. IEEE Trans Circuits Syst II Analog Digit Signal Process 40(3):163–173
16. Strogatz S, Abrams DM, McRobie A et al (2005) Theoretical mechanics: crowd synchrony on the millennium bridge. Nature 438(7064):43–44
17. Tass PA (1999) Phase resetting in medicine and biology. Springer, Berlin
18. Wang W, Kiss IZ, Hudson JL (2001) Clustering of arrays of chaotic chemical oscillators by feedback and forcing. Phys Rev Lett 86(21):4954
19. Wiesenfeld K, Colet P, Strogatz SH (1998) Frequency locking in Josephson arrays: connection with the Kuramoto model. Phys Rev E 57(2):1563
20. Wikipedia (2017) Wireless sensor network—Wikipedia, the free encyclopedia
21. Winfree A (1980) The geometry of biological time. Springer, New York

Verification of Biomedical Processes with Anomalous Diffusion, Transport and Interaction of Species

Messoud Efendiev and Vitali Vougalter

Abstract The paper deals with the easily verifiable necessary condition of the preservation of the nonnegativity of the solutions of a system of parabolic equations in the case of the anomalous diffusion with the Laplace operator in a fractional power in one dimension. This necessary condition is vitally important for the applied analysis society because it imposes the necessary form of the system of equations that must be studied mathematically.

Keywords Anomalous diffusion · Parabolic systems · Nonnegativity of solutions

AMS Subject Classification 35K55 · 35K57

1 Introduction

The solutions of many systems of convection-diffusion-reaction equations arising in biology, physics or engineering describe such quantities as population densities, pressure or concentrations of nutrients and chemicals. Thus, a natural property to require for the solutions is the nonnegativity. Models that do not guarantee the nonnegativity are not valid or break down for small values of the solution. In many cases, showing that a particular model does not preserve the nonnegativity leads to the better understanding of the model and its limitations. One of the first steps in analyzing ecological or biological or bio-medical models mathematically is to test whether solutions originating from the nonnegative initial data remain nonnegative

M. Efendiev (✉)
Helmholtz Zentrum München, Institut Für Computational Biology, Ingolstädter Landstrasse 1, 85764 Neunerberg, Germany
e-mail: messoud.efendiyev@helmholtz-muenchen.de

Department of Mathematics, Marmara University, Istanbul, Turkey

V. Vougalter
Department of Mathematics, University of Toronto, Toronto, ON M5S 2E4, Canada
e-mail: vitali@math.toronto.edu

© Higher Education Press 2021
D. Volchenkov (ed.), *Nonlinear Dynamics, Chaos, and Complexity*,
Nonlinear Physical Science, https://doi.org/10.1007/978-981-15-9034-4_5

(as long as they exist). In other words, the model under consideration ensures that the nonnegative cone is positively invariant. We recall that if the solutions (of a given evolution PDE) corresponding to the nonnegative initial data remain nonnegative as long as they exist, we say that the system satisfies the nonnegativity property.

For scalar equations the nonnegativity property is a direct consequence of the maximum principle (see [2] and the references therein). However, for systems of equations the maximum principle is not valid. In the particular case of monotone systems the situation resembles the case of scalar equations, sufficient conditions for preserving the nonnegative cone can be found in [2].

In this paper we aim to prove a simple and easily verifiable criterion, that is, the necessary condition for the nonnegativity of solutions of systems of nonlinear convection-anomalous diffusion-reaction equations arising in the modelling of life sciences. We believe that it could provide the modeler with a tool, which is easy to verify, to approach the question of positive invariance of the model.

The present article deals with the preservation of the nonnegativity of solutions of the following system of reaction-diffusion equations

$$\frac{\partial u}{\partial t} = -A(-\Delta_x)^s u + \sum_{l=1}^{m} \Gamma^l \frac{\partial u}{\partial x_l} - F(u), \tag{1}$$

where A, Γ^l, $1 \leq l \leq m$ are $N \times N$ matrices with constant coefficients, which is relevant to the cell population dynamics in Mathematical Biology. We call system (1) as a (N, m) one. Note that the analogous model can be used to study such branches of science as the Damage Mechanics, the temperature distribution in Thermodynamics. In the present article the space variable x corresponds to the cell genotype, $u_k(x, t)$ stands for the cell density distributions for various groups of cells as functions of their genotype and time,

$$u(x, t) = (u_1(x, t), u_2(x, t), \ldots, u_N(x, t))^T.$$

The operator $(-\Delta_x)^s$ in problem (1) describes a particular case of anomalous diffusion actively treated in the context of different applications in plasma physics and turbulence [1, 8], surface diffusion [4, 6], semiconductors [7] and so on. Anomalous diffusion can be described as a random process of particle motion characterized by the probability density distribution of jump length. The moments of this density distribution are finite in the case of normal diffusion, but this is not the case for superdiffusion. Asymptotic behavior at infinity of the probability density function determines the value s of the power of the Laplacian [5]. The operator $(-\Delta_x)^s$ is defined by virtue of the spectral calculus. For the simplicity of presentation we will treat the case of the one spatial dimension with $0 < s < 1/4$. Front propagation problems with anomalous diffusion were studied actively in recent years (see e.g. [9, 10]). The solvability of the single equation containing the Laplacian with drift relevant to the fluid mechanics was treated in [11]. We assume here that (1) contains the square matrices with the entries constant in space and time

$$(A)_{k,j} := a_{k,j}, \quad (\Gamma)_{k,j} := \gamma_{k,j}, \quad 1 \le k, j \le N$$

and that the matrix $A + A^* > 0$ for the sake of the global well posedness of system (1). Here A^* stands for the adjoint of matrix A. Hence, problem (1) can be rewritten in the form

$$\frac{\partial u_k}{\partial t} = -\sum_{j=1}^{N} a_{k,j} \left(-\frac{\partial^2}{\partial x^2} \right)^s u_j + \sum_{j=1}^{N} \gamma_{k,j} \frac{\partial u_j}{\partial x} - F_k(u), \quad 1 \le k \le N, \quad (2)$$

with $0 < s < \dfrac{1}{4}$. In the present work the interaction of species term

$$F(u) = (F_1(u), F_2(u), \ldots, F_N(u))^T,$$

which can be linear, nonlinear or in principle even nonlocal. We assume its smoothness in the theorem below for the sake of the well posedness of our system (1), although, we are not focused on the well posedness issue in the present article. Let us choose the space dimension $d = 1$, which is related to the solvability conditions for the linear Poisson type problem (14) stated in Lemma 2 below. From the perspective of applications, the space dimension is not restricted to $d = 1$ because the space variable corresponds to cell genotype but not to the usual physical space. We denote the inner product as

$$(f(x), g(x))_{L^2(\mathbb{R})} := \int_{-\infty}^{\infty} f(x)\bar{g}(x)dx, \quad (3)$$

with a slight abuse of notations when the functions involved in (3) are not square integrable, like for example the one present in orthogonality relations (17) and (18) of Lemma 2 below. Indeed, if $f(x) \in L^1(\mathbb{R})$ and $g(x)$ is bounded, then the integral in the right side of (3) makes sense. As for the vector functions, their inner product is defined using their components as

$$(u, v)_{L^2(\mathbb{R}, \mathbb{R}^N)} := \sum_{k=1}^{N} (u_k, v_k)_{L^2(\mathbb{R})}. \quad (4)$$

Clearly, (4) induces the norm

$$\|u\|^2_{L^2(\mathbb{R}, \mathbb{R}^N)} = \sum_{k=1}^{N} \|u_k\|^2_{L^2(\mathbb{R})}.$$

We use the Sobolev spaces

$$H^{2s}(\mathbb{R}) := \left\{ u(x) : \mathbb{R} \to \mathbb{R} \mid u(x) \in L^2(\mathbb{R}), \; \left(-\frac{d^2}{dx^2} \right)^s u \in L^2(\mathbb{R}) \right\}, \quad 0 < s \leq 1$$

equipped with the norm

$$\|u\|^2_{H^{2s}(\mathbb{R})} := \|u\|^2_{L^2(\mathbb{R})} + \left\| \left(-\frac{d^2}{dx^2} \right)^s u \right\|^2_{L^2(\mathbb{R})}. \tag{5}$$

By the nonnegativity of a vector function below we mean the nonnegativity of the each of its components. Our main statement is as follows.

Theorem 1 *Let $F : \mathbb{R}^N \to \mathbb{R}^N$, such that $F \in C^1$, the initial condition for system (1) is $u(x, 0) = u_0(x) \geq 0$ and $u_0(x) \in L^2(\mathbb{R}, \mathbb{R}^N)$. We also assume that the off diagonal element of the matrix A, are nonnegative, such that*

$$a_{k,l} \geq 0, \quad 1 \leq k, l \leq N, \quad k \neq l. \tag{6}$$

Then the necessary condition for problem (1) to admit a solution $u(x, t) \geq 0$ for all $t \in [0, \infty)$ is that the matrices A and Γ are diagonal and for all $1 \leq k \leq N$

$$F_k(s_1, \ldots, s_{k-1}, 0, s_{k+1}, \ldots, s_N) \leq 0 \tag{7}$$

holds, where $s_l \geq 0$ and $1 \leq l \leq N$, $l \neq k$.

Remark 1 In the case of the linear interaction of species, namely when $F(u) = Lu$, where L is a matrix with elements $b_{i,j}$, $1 \leq i, j \leq N$ constant in space and time, our necessary condition leads to the condition that the matrix L must be essentially nonpositive, that is $b_{i,j} \leq 0$ for $i \neq j$, $1 \leq i, j \leq N$.

Remark 2 Our proof implies that, the necessary condition for preserving the non-negative cone is carried over from the ODE (the spatially homogeneous case, as described by the ordinary differential equation $u'(t) = -F(u)$) to the case of the anomalous diffusion and the convective drift term.

Remark 3 In the forthcoming papers we intend to consider the following cases:

(a) the necessary and sufficient conditions of the present work,
(b) the nonautonomous version of the present work,
(c) the density-dependent diffusion matrix,
(d) the effect of the delay term in the cases (a), (b) and (c).

Let us proceed to the proof of our main result.

2 The Preservation of the Nonnegativity of the Solution of the System of Parabolic Equations

Proof of Theorem 1 Let us note that the maximum principle actively used for the studies of solutions of single parabolic equations does not apply to systems of such equations. We consider a time independent, square integrable vector function $v(x)$ and estimate

$$\left(\frac{\partial u}{\partial t}\Big|_{t=0}, v\right)_{L^2(\mathbb{R},\mathbb{R}^N)} = \left(\lim_{t\to 0}\frac{u(x,t) - u_0(x)}{t}, v(x)\right)_{L^2(\mathbb{R},\mathbb{R}^N)}.$$

By means of the continuity of the inner product, the right side of the identity above equals to

$$\lim_{t\to 0}\frac{(u(x,t), v(x))_{L^2(\mathbb{R},\mathbb{R}^N)}}{t} - \lim_{t\to 0}\frac{(u_0(x), v(x))_{L^2(\mathbb{R},\mathbb{R}^N)}}{t}. \tag{8}$$

Let us choose the initial condition for our system $u_0(x) \geq 0$ and the constant in time vector function $v(x) \geq 0$ to be orthogonal to each other in $L^2(\mathbb{R}, \mathbb{R}^N)$. This can be achieved, for instance for

$$u_0(x) = (\tilde{u}_1(x), \ldots, \tilde{u}_{k-1}(x), 0, \tilde{u}_{k+1}(x), \ldots, \tilde{u}_N(x)), \quad v_j(x) = \tilde{v}(x)\delta_{j,k}, \tag{9}$$

with $1 \leq j \leq N$, where $\delta_{j,k}$ is the Kronecker symbol and $1 \leq k \leq N$ is fixed. Therefore, the second term in (8) vanishes and (8) equals to

$$\lim_{t\to 0}\frac{\sum_{k=1}^{N}\int_{-\infty}^{\infty} u_k(x,t)v_k(x)dx}{t} \geq 0$$

due to the nonnegativity of all the components $u_k(x, t)$ and $v_k(x)$ involved in the formula above. Thus, we arrive at

$$\sum_{j=1}^{N}\int_{-\infty}^{\infty}\frac{\partial u_j}{\partial t}\Big|_{t=0} v_j(x)dx \geq 0.$$

By virtue of (9), only the kth component of the vector function $v(x)$ is nontrivial. This yields

$$\int_{-\infty}^{\infty}\frac{\partial u_k}{\partial t}\Big|_{t=0} \tilde{v}(x)dx \geq 0.$$

Hence, via (2) we arrive at

$$\int_{-\infty}^{\infty}\left[-\sum_{j=1,\ j\neq k}^{N}a_{k,j}\left(-\frac{\partial^2}{\partial x^2}\right)^s\tilde{u}_j(x)+\sum_{j=1,\ j\neq k}^{N}\gamma_{k,j}\frac{\partial\tilde{u}_j}{\partial x}-\right.$$

$$\left.-F_k(\tilde{u}_1(x),\ldots,\tilde{u}_{k-1}(x),0,\tilde{u}_{k+1}(x),\ldots,\tilde{u}_N(x))\right]\tilde{v}(x)dx\geq 0.$$

Since the nonnegative, square integrable function $\tilde{v}(x)$ can be chosen arbitrarily, we obtain

$$-\sum_{j=1,\ j\neq k}^{N}a_{k,j}\left(-\frac{\partial^2}{\partial x^2}\right)^s\tilde{u}_j(x)+\sum_{j=1,\ j\neq k}^{N}\gamma_{k,j}\frac{\partial\tilde{u}_j}{\partial x}-$$

$$-F_k(\tilde{u}_1(x),\ldots,\tilde{u}_{k-1}(x),0,\tilde{u}_{k+1}(x),\ldots,\tilde{u}_N(x))\geq 0\quad a.e.\qquad(10)$$

For the purpose of the scaling, let us replace all the $\tilde{u}_j(x)$ by $\tilde{u}_j\left(\dfrac{x}{\varepsilon}\right)$ in the inequality above, where $\varepsilon>0$ is a small parameter. This yields

$$-\sum_{j=1,\ j\neq k}^{N}\frac{a_{k,j}}{\varepsilon^{2s}}\left(-\frac{\partial^2}{\partial y^2}\right)^s\tilde{u}_j(y)+\sum_{j=1,\ j\neq k}^{N}\frac{\gamma_{k,j}}{\varepsilon}\frac{\partial\tilde{u}_j(y)}{\partial y}-$$

$$-F_k(\tilde{u}_1(y),\ldots,\tilde{u}_{k-1}(y),0,\tilde{u}_{k+1}(y),\ldots,\tilde{u}_N(y))\geq 0\quad a.e.\qquad(11)$$

Clearly, the second term in the left side of (11) is the leading one as $\varepsilon\to 0$. In the case of $\gamma_{k,j}>0$ we can choose here $\tilde{u}_j(y)=e^{-y}$ in a neighborhood of the origin, smooth and decaying to zero at the infinities. If $\gamma_{k,j}<0$, then we can pick $\tilde{u}_j(y)=e^y$ around the origin and tending to zero at the infinities. Then the left side of (11) can be made as negative as possible which will violate inequality (11). Note that the last term in the left side of (11) will remain bounded. Therefore, for the matrix Γ involved in problem (1), the off diagonal terms should vanish, such that

$$\gamma_{k,j}=0,\quad 1\leq k,j\leq N,\quad k\neq j.$$

Therefore, from (11) we obtain

$$-\sum_{j=1,\ j\neq k}^{N}\frac{a_{k,j}}{\varepsilon^{2s}}\left(-\frac{\partial^2}{\partial y^2}\right)^s\tilde{u}_j(y)-$$

$$-F_k(\tilde{u}_1(y),\ldots,\tilde{u}_{k-1}(y),0,\tilde{u}_{k+1}(y),\ldots,\tilde{u}_N(y))\geq 0\quad a.e.\qquad(12)$$

Let us suppose that some of the $a_{k,j}$ involved in the sum in the left side of (12) are strictly positive. We choose here all the $\tilde{u}_j(y)$, $1\leq j\leq N$, $j\neq k$ to be identical. For the equation

$$-\left(-\frac{\partial^2}{\partial x^2}\right)^s \tilde{u}_j(x) = \tilde{v}_j(x), \quad 0 < s < \frac{1}{4}, \tag{13}$$

we assume that its right side belongs to $C_c^\infty(\mathbb{R})$. Clearly, $\tilde{v}_j(x) \in L^1(\mathbb{R}) \cap L^2(\mathbb{R})$ as well. Then by means of the part 1 of Lemma 2 below, (13) admits a unique solution $\tilde{u}_j(x) \in H^{2s}(\mathbb{R})$. Suppose the right side of (13) is nonnegative on the whole real line. By virtue of Sect. 5.9 of [3] we have the explicit formula

$$\tilde{u}_j(x) = -c_s \int_{-\infty}^{\infty} |x - y|^{2s-1} \tilde{v}_j(y) dy,$$

where $c_s > 0$ is a constant. Then $\tilde{u}_j(x)$ is negative on $\mathbb{R}$, which contradicts to our original assumption. Therefore, $\tilde{v}_j(x)$ has the points of negativity on the real line. By making the parameter ε small enough, we can violate the inequality in (12). Since the negativity of the off diagonal elements of the matrix A is ruled out due to assumption (6), we arrive at

$$a_{k,j} = 0, \quad 1 \le k, j \le N, \quad k \ne j.$$

Therefore, by means of (10) we obtain

$$F_k(\tilde{u}_1(x), \ldots, \tilde{u}_{k-1}(x), 0, \tilde{u}_{k+1}(x), \ldots, \tilde{u}_N(x)) \le 0 \quad a.e.,$$

where $\tilde{u}_j(x) \ge 0$ and $\tilde{u}_j(x) \in L^2(\mathbb{R})$ with $1 \le j \le N, \ j \ne k$. ∎

3　Auxiliary Results

Below we state the solvability conditions for the linear Poisson type equation with a square integrable right side

$$\left(-\frac{d^2}{dx^2}\right)^s u = f(x), \quad x \in \mathbb{R}, \quad 0 < s < 1. \tag{14}$$

We have the following technical proposition. It can be easily derived by applying the standard Fourier transform

$$\widehat{\phi}(p) := \frac{1}{\sqrt{2\pi}} \int_{-\infty}^{\infty} \phi(x) e^{-ipx} dx. \tag{15}$$

to both sides of problem (14) (see Lemma 1.6 of [13]). For the similar results in three dimensions see Lemma 5 of [12]. We will use the obvious upper bound

$$\|\widehat{\phi}(p)\|_{L^\infty(\mathbb{R})} \le \frac{1}{\sqrt{2\pi}} \|\phi(x)\|_{L^1(\mathbb{R})}. \tag{16}$$

We will provide the proof below for the convenience of the readers.

Lemma 1 *Let $f(x) : \mathbb{R} \to \mathbb{R}$ and $f(x) \in L^2(\mathbb{R})$.*

(1) When $0 < s < \frac{1}{4}$ and in addition $f(x) \in L^1(\mathbb{R})$, Eq. (14) admits a unique solution $u(x) \in H^{2s}(\mathbb{R})$.

(2) When $\frac{1}{4} \leq s < \frac{3}{4}$ and additionally $|x| f(x) \in L^1(\mathbb{R})$, problem (14) possesses a unique solution $u(x) \in H^{2s}(\mathbb{R})$ if and only if the orthogonality relation

$$(f(x), 1)_{L^2(\mathbb{R})} = 0 \tag{17}$$

holds.

(3) When $\frac{3}{4} \leq s < 1$ and in addition $x^2 f(x) \in L^1(\mathbb{R})$, Eq. (14) has a unique solution $u(x) \in H^{2s}(\mathbb{R})$ if and only if orthogonality conditions (17) and

$$(f(x), x)_{L^2(\mathbb{R})} = 0 \tag{18}$$

hold.

Proof First, let us observe that by virtue of norm definition (5) along with the square integrability of the right side of (14), it would be sufficient to establish the solvability of Eq. (14) in $L^2(\mathbb{R})$. Clearly, the solution $u(x) \in L^2(\mathbb{R})$ will belong to $H^{2s}(\mathbb{R})$, $0 < s < 1$ as well.

We prove the uniqueness of solutions for problem (14). If $u_{1,2}(x) \in H^{2s}(\mathbb{R})$ both solve (14), then the difference $w(x) := u_1(x) - u_2(x) \in L^2(\mathbb{R})$ satisfies the homogeneous equation

$$\left(-\frac{d^2}{dx^2} \right)^s w = 0.$$

Because the operator $\left(-\frac{d^2}{dx^2} \right)^s$ on the real line does not possess nontrivial square integrable zero modes, $w(x) = 0$ a.e. on $\mathbb{R}$.

We apply (15) to both sides of problem (14). This yields

$$\widehat{u}(p) = \frac{\widehat{f}(p)}{|p|^{2s}} \chi_{\{p \in \mathbb{R} \mid |p| \leq 1\}} + \frac{\widehat{f}(p)}{|p|^{2s}} \chi_{\{p \in \mathbb{R} \mid |p| > 1\}}, \tag{19}$$

where χ_A is the characteristic function of a set $A \subseteq \mathbb{R}$. Evidently, for all $0 < s < 1$ the second term in the right side of (19) is square integrable by means of the bound

$$\int_{-\infty}^{\infty} \frac{|\widehat{f}(p)|^2}{|p|^{4s}} \chi_{\{p \in \mathbb{R} \mid |p| > 1\}} dp \leq \|f\|^2_{L^2(\mathbb{R})} < \infty.$$

To establish the square integrability of the first term in the right side of (19) for $0 < s < \frac{1}{4}$, we apply inequality (16), which yields

$$\int_{-\infty}^{\infty} \frac{|\widehat{f}(p)|^2}{|p|^{4s}} \chi_{\{p\in\mathbb{R} \ | \ |p|\leq 1\}} dp \leq \frac{\|f(x)\|_{L^1(\mathbb{R})}^2}{\pi(1-4s)} < \infty.$$

This completes the proof of part (1) of our lemma.

To prove the solvability of problem (14) when $\frac{1}{4} \leq s < \frac{3}{4}$, we apply the formula

$$\widehat{f}(p) = \widehat{f}(0) + \int_0^p \frac{d\widehat{f}(s)}{ds} ds.$$

This enables us to express the first term in the right side of (19) as

$$\frac{\widehat{f}(0)}{|p|^{2s}} \chi_{\{p\in\mathbb{R} \ | \ |p|\leq 1\}} + \frac{\int_0^p \frac{d\widehat{f}(s)}{ds} ds}{|p|^{2s}} \chi_{\{p\in\mathbb{R} \ | \ |p|\leq 1\}}. \tag{20}$$

By means of definition (15)

$$\left|\frac{d\widehat{f}(p)}{dp}\right| \leq \frac{1}{\sqrt{2\pi}} \||x|f(x)\|_{L^1(\mathbb{R})} < \infty$$

via the one of our assumptions. Thus,

$$\left|\frac{\int_0^p \frac{d\widehat{f}(s)}{ds} ds}{|p|^{2s}} \chi_{\{p\in\mathbb{R} \ | \ |p|\leq 1\}}\right| \leq \frac{1}{\sqrt{2\pi}} \||x|f(x)\|_{L^1(\mathbb{R})} |p|^{1-2s} \chi_{\{p\in\mathbb{R} \ | \ |p|\leq 1\}} \in L^2(\mathbb{R}).$$

The remaining term in (20) $\dfrac{\widehat{f}(0)}{|p|^{2s}} \chi_{\{p\in\mathbb{R} \ | \ |p|\leq 1\}} \in L^2(\mathbb{R})$ if and only if $\widehat{f}(0) = 0$, which gives us orthogonality relation (17) in case (2) of the lemma.

Finally, it remains to study the situation when $\dfrac{3}{4} \leq s < 1$. For that purpose, we use the identity

$$\widehat{f}(p) = \widehat{f}(0) + p\frac{d\widehat{f}}{dp}(0) + \int_0^p \left(\int_0^r \frac{d^2\widehat{f}(q)}{dq^2} dq\right) dr.$$

This allows us to express the first term in the right side of (19) as

$$\left[\frac{\widehat{f}(0)}{|p|^{2s}} + \frac{p\frac{d\widehat{f}}{dp}(0)}{|p|^{2s}} + \frac{\int_0^p \left(\int_0^r \frac{d^2\widehat{f}(q)}{dq^2} dq\right) dr}{|p|^{2s}}\right] \chi_{\{p\in\mathbb{R} \ | \ |p|\leq 1\}}. \tag{21}$$

Definition (15) yields

$$\left|\frac{d^2\widehat{f}(p)}{dp^2}\right| \leq \frac{1}{\sqrt{2\pi}} \|x^2 f(x)\|_{L^1(\mathbb{R})} < \infty$$

as assumed. This enables us to estimate

$$\left| \frac{\int_0^p \left(\int_0^r \frac{d^2 \widehat{f}(q)}{dq^2} dq \right) dr}{|p|^{2s}} \chi_{\{p \in \mathbb{R} \mid |p| \leq 1\}} \right| \leq \frac{1}{2\sqrt{2\pi}} \|x^2 f(x)\|_{L^1(\mathbb{R})} |p|^{2-2s} \chi_{\{p \in \mathbb{R} \mid |p| \leq 1\}},$$

which is clearly square integrable. The sum of the first and the second terms in (21) does not belong to $L^2(\mathbb{R})$ unless both $\widehat{f}(0)$ and $\frac{d\widehat{f}}{dp}(0)$ are equal to zero. This yields orthogonality relations (17) and (18) respectively. ∎

Let us note that the left side of relations (17) and (18) is well defined under the given conditions. For the lower values of the power of the negative second derivative operator $0 < s < \frac{1}{4}$ under the assumptions stated above no orthogonality relations are required to solve the linear Poisson type equation (14) in $H^{2s}(\mathbb{R})$.

References

1. Carreras B, Lynch V, Zaslavsky G (2001) Anomalous diffusion and exit time distribution of particle tracers in plasma turbulence model. Phys Plasmas 8:5096–5103
2. Efendiev MA (2013) Evolution equations arising in the modelling of life sciences. International series of numerical mathematics, vol 163. Birkhäuser/Springer, Basel, 217 pp
3. Lieb E, Loss M (1997) Analysis. Graduate studies in mathematics, vol 14. American Mathematical Society, Providence
4. Manandhar P, Jang J, Schatz GC, Ratner MA, Hong S (2003) Anomalous surface diffusion in nanoscale direct deposition processes. Phys Rev Lett 90:4043–4052
5. Metzler R, Klafter J (2000) The random walk's guide to anomalous diffusion: a fractional dynamics approach. Phys Rep 339:1–77
6. Sancho J, Lacasta A, Lindenberg K, Sokolov I, Romero A (2004) Diffusion on a solid surface: Anomalous is normal. Phys Rev Lett 92:250601
7. Scher H, Montroll E (1975) Anomalous transit-time dispersion in amorphous solids. Phys Rev B 12:2455–2477
8. Solomon T, Weeks E, Swinney H (1993) Observation of anomalous diffusion and Lévy flights in a two-dimensional rotating flow. Phys Rev Lett 71:3975–3978
9. Volpert VA, Nec Y, Nepomnyashchy AA (2010) Exact solutions in front propagation problems with superdiffusion. Phys D 239(3–4):134–144
10. Volpert VA, Nec Y, Nepomnyashchy AA (2013) Fronts in anomalous diffusion-reaction systems. Philos Trans R Soc Lond Ser A Math Phys Eng Sci 371(1982):20120179, 18pp
11. Vougalter V, Volpert V (2012) On the solvability conditions for the diffusion equation with convection terms. Commun Pure Appl Anal 11(1):365–373
12. Vougalter V, Volpert V (2015) Existence of stationary solutions for some integro-differential equations with anomalous diffusion. J Pseudo-Differ Oper Appl 6(4):487–501
13. Vougalter V, Volpert V (2018) Solvability of some integro-differential equations with anomalous diffusion. In: Regularity and stochasticity of nonlinear dynamical systems. Nonlinear systems and complexity, vol 21. Springer, Cham, pp 1–17

Chaos-Based Communication Using Isochronal Synchronization: Considerations About the Synchronization Manifold

J. M. V. Grzybowski, E. E. N. Macau, and T. Yoneyama

Abstract In this work, it is evaluated the conditions for the existence of a synchronization manifold in coupled systems. The objective is to enhance the comprehension of the possibilities and restrictions that are involved in the implementation of simultaneous bidirectional communication schemes. As a result, general conditions that rule the existence of the set $\mathcal{S}$ for the commonly appearing coupling designs belonging to the context of chaos-based communication are unveiled.

1 Introduction

Under favourable conditions, delay-coupled chaotic oscillators can settle down into a stable zero-lag synchronous behavior, known as isochronal synchronization. Such synchronous state has been explored for chaos-based communication in a variety of schemes using lasers and electronic circuits [1, 7, 15, 17, 18]. One necessary condition for such realization is the existence of the synchronization manifold, which is the set in which the synchronous solution evolves, frequently defined as $\mathcal{S} = \{x_i(t) \in \mathbb{R}^n,\ i = 1, ..., N;\ x_1(t) = \cdots = x_N(t)\}$ for a network with N oscillators, where $x_i(t)$ are state vectors of its nodes. [10–12]. It is interesting to note that not all coupling schemes and network topologies allow the existence of such a set. In the context of chaos-based communication, this restricts the choice of parameters in

J. M. V. Grzybowski
Federal University of Fronteira Sul, Rod. 135 km 72, Erechim, RS, Brazil
e-mail: jose.grzybowski@uffs.edu.br

E. E. N. Macau (✉)
Federal University of São Paulo, Av. Cesare Monsueto Giulio Lattes,
São José dos Campos, SP 1211, Brazil
e-mail: elbert.macau@inpe.br

T. Yoneyama
Aeronautics Institute of Technology, Pa. Mal. Eduardo Gomes, 50,
São José dos Campos, SP, Brazil
e-mail: takashi@ita.br

© Higher Education Press 2021
D. Volchenkov (ed.), *Nonlinear Dynamics, Chaos, and Complexity*,
Nonlinear Physical Science, https://doi.org/10.1007/978-981-15-9034-4_6

the coupling setup, which has to take into account the conditions for the existence of the synchronization manifold S as prerequisite for the design of communication schemes.

Considering the interest of recent publications in the implementation of chaos-based communication schemes using isochronal synchronization, this paper evaluates the conditions for the existence of a synchronization manifold in recently studied coupling setups [1, 8, 9, 14–18]. The objective is to enhance the comprehension of the general conditions that are involved in the implementation of simultaneous bidirectional communication schemes, such as those presented in the studies [1, 15, 17, 18], in what concerns the existence of a synchronous solution. As a result, general conditions that rule the existence of the set S for the commonly appearing coupling designs belonging to the context of chaos-based communication are unveiled.

2   Chaos-Based Communication and Coupling Schemes for Isochronal Synchronization

Isochronal synchronization is estimated to bear simultaneous bidirectional communication at bit rates higher than 10 Gbps [6], while also allows the negotiation of secret keys through public communication channels [7]. These features have attracted the attention of researchers for its application in chaos-based communication schemes [1, 7, 15, 17, 18]. At the current stage of development of such schemes, efforts are partly devoted to the understanding of the mechanisms behind the emergence and maintenance of isochronal synchronous state, including considerations about the existence of the synchronization manifold [11, 12].

On the other hand, delayed coupling isochronal synchronization is highly unstable. As so, a number of techniques have been developed and proposed that are aimed at stabilizing stable isochronal synchronous behavior among systems in chaotic regime [9, 14, 15, 18]. While these approaches work, they frequently require that strict symmetry conditions be met, which often make practical applications more demanding in respect of their design and constituting elements.

Regardless of these practical difficulties, synchronization stability issues have been addressed using existing techniques which mainly consider the use of delayed [15] and direct [2] self-feedback or the inclusion of a third oscillator to induce the injection-locking effect [18]. As a complement to such studies, in this work we present a number of theoretical considerations about the existence of a synchronization. We cover several topologies of communication based on isochronal synchronization of chaos and we explore conditions under which the synchronization manifold exists.

2.1 Delayed Coupling and Delayed Self-coupling

Fischer et al. [15] studied simultaneous bidirectional transmission using coupled semiconductor lasers (SLs) operating in chaotic regime. To avoid LLS (leader-laggard synchronization) and induce stable IS (isochronal synchronization), the mutual coupling of the SLs was combined with optical delayed self-feedback by means of a partially transparent mirror positioned in the communication channel between the lasers, as shown in Fig. 1. The coupling signal in this scheme is a combination of the delayed outputs of both lasers, where delays are caused by propagation time and delay magnitude are larger than the time scale of the relaxation oscillations. It is remarked that synchronization establishes for arbitrary distances between the lasers [15], which is highly desirable from the communication viewpoint, since it means that the phenomenology of isochronal synchronization is not a barrier for long-distance communication. Nevertheless, it must also be recognized that other limiting technical factors for long-distance communication certainly exist (Fig. 2).

To derive the conditions for the existence of the synchronization manifold, consider the equations representing the coupled system as

$$\dot{x}_1(t) = Ax_1(t) + g\,(x_1(t)) + K_{11}x_1(t - \tau_{11}) + K_{21}x_2(t - \tau_{21}) \qquad (1)$$

$$\dot{x}_2(t) = Ax_2(t) + g\,(x_2(t)) + K_{22}x_2(t - \tau_{22}) + K_{12}x_1(t - \tau_{12}) \qquad (2)$$

where $x_1,\ x_2 \in \mathbb{R}^n$. Define

$$x(t) = \begin{bmatrix} x_1(t) \\ x_2(t) \end{bmatrix} \qquad (3)$$

$$M = \begin{bmatrix} 1 & -1 \end{bmatrix} \otimes I_n \qquad (4)$$

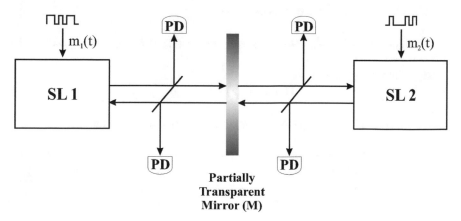

Fig. 1 Mutually coupled semiconductor lasers with optical self-feedback due to effect of a partially transparent mirror: for this configuration, isochronal synchronization can be obtained for arbitrary distances between the lasers [15]. Photo-Diodes (PD) detect the outputs of SL1 and SL2; binary messages are simultaneously injected in SL1 and SL2 at rates of 1 Gpbs and recovered at the other end

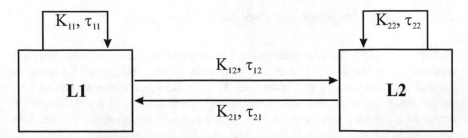

Fig. 2 Two mutually delay-coupled systems featuring delayed self-feedback (τ_{11}, $\tau_{22} \neq 0$): systems receive delayed state information from each other and delayed self-feedback

and the synchronization manifold can be written as

$$S = \{x(t), \ Mx(t) = 0\} \tag{5}$$

Further, define the error system as

$$\begin{aligned}
\dot{x}_1(t) - \dot{x}_2(t) = & Ax_1(t) - Ax_2(t) + g\,(x_1(t)) - g\,(x_2(t)) + K_{11}x_1(t - \tau_{11}) \\
& - K_{22}x_2(t - \tau_{22}) + K_{21}x_2(t - \tau_{21}) - K_{12}x_1(t - \tau_{12})
\end{aligned} \tag{6}$$

and suppose $Mx(t) = 0$ holds, which implies $x_1(t) = x_2(t) = s(t)$, such that Eq. 6 can be rewritten as

$$\begin{aligned}
\dot{s}(t) - \dot{s}(t) = & As(t) - As(t) + g\,(s(t)) - g\,(s(t)) + K_{11}s(t - \tau_{11}) \\
& - K_{22}s(t - \tau_{22}) + K_{21}s(t - \tau_{21}) - K_{12}s(t - \tau_{12})
\end{aligned} \tag{7}$$

It follows that the existence of S is conditioned to

$$K_{11}s(t - \tau_{11}) - K_{22}s(t - \tau_{22}) + K_{21}s(t - \tau_{21}) - K_{12}s(t - \tau_{12}) = 0 \tag{8}$$

Regarding the system

$$\dot{s}(t) = As(t) + g\,(s(t)) \tag{9}$$

as chaotic, it follows that Eq. 8 holds under any of the conditions below:

$$\begin{aligned}
& \{K_{11} = -K_{21}, \ K_{22} = -K_{12}, \ \tau_{11} = \tau_{21}, \ \tau_{22} = \tau_{12} \\
& \{K_{11} = -K_{21}, \ K_{22} = -K_{12}, \ \tau_{11} = \tau_{21}, \ \tau_{22} = \tau_{12} \\
& \{K_{11} = -K_{21}, \ K_{22} = -K_{12}, \ \tau_{11} = \tau_{21}, \ \tau_{22} = \tau_{12} \\
& \{K_{11} + K_{21} = K_{12} + K_{22}, \ \tau_{11} = \tau_{12} = \tau_{21} = \tau_{22}
\end{aligned} \tag{10}$$

2.2 Injection Locking and Mutual Delayed Coupling

Consider the coupling setup presented in Fig. 3, and write the equations of the coupled oscillators as

$$\dot{x}_1(t) = Ax_1(t) + g(x_1(t)) + K_{21}x_2(t - \tau_{21}) + K_{31}x_3(t - \tau_{31}) \qquad (11)$$

$$\dot{x}_2(t) = Ax_2(t) + g(x_2(t)) + K_{12}x_1(t - \tau_{12}) + K_{32}x_3(t - \tau_{32}) \qquad (12)$$

$$\dot{x}_3(t) = Bx_3(t) + h(x_3(t)) \qquad (13)$$

As isochronal synchronization between systems 1 and 2 is aimed, the error system can be defined as $e = x_1(t) - x_2(t)$, which yields

$$\begin{aligned} \dot{x}_1 - \dot{x}_2 = {} & Ax_1(t) - Ax_2(t) + g(x_1(t)) - g(x_2(t)) + K_{21}x_2(t - \tau_{21}) \\ & - K_{12}x_1(t - \tau_{12}) + K_{31}x_3(t - \tau_{31}) - K_{32}x_3(t - \tau_{32}) \end{aligned} \qquad (14)$$

Define

$$x(t) = \begin{bmatrix} x_1(t) \\ x_2(t) \end{bmatrix} \qquad (15)$$

$$M = \begin{bmatrix} 1 & -1 \end{bmatrix} \otimes I_n \qquad (16)$$

and the synchronization manifold can be defined as in Eq. 5. Suppose that $Mx(t) = 0$ holds, then the error system 14 must allow

$$K_{21}s(t - \tau_{21}) - K_{12}s(t - \tau_{12}) + K_{31}s(t - \tau_{31}) - K_{32}s(t - \tau_{32}) = 0 \qquad (17)$$

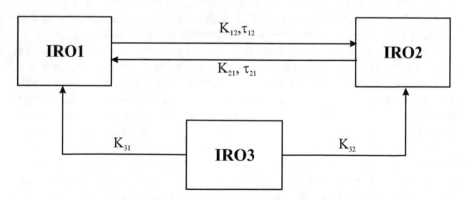

Fig. 3 Ikeda ring oscillators with round-trip time τ_R where IRO1 and IRO2 are mutually coupled: isochronal synchronization between IRO1 and IRO2 is obtained by the unidirectional drive performed by IRO3 in both systems simultaneously. For isochronal synchronization to emerge, the delays from IRO3 to IRO1 and IRO2 must be the same, that is, $\tau_{31} = \tau_{32}$

Suppose the system

$$\dot{s}(t) = As(t) + g\,(s(t)) \tag{18}$$

is chaotic, then Eq. 17 holds if and only if

$$K_{31} = K_{32}, \; K_{21} = K_{12}, \; \tau_{31} = \tau_{32}, \; \tau_{21} = \tau_{12} \tag{19}$$

Note that, from the communication viewpoint, the need for a third oscillator is inconvenient, since it makes the scheme more resource demanding than two-oscillator schemes. Besides demanding a third oscillator, it would also demand a channel and bandwidth for the transmission of the synchronization signal from IRO3 to IRO1 and IRO2 during operation. Furthermore, a direct consequence of the presence of IRO3 is that IS between IRO1 and to IRO2 depends on the symmetry of delays from the signals coming from IRO3, i.e., it is required that $\tau_{31} = \tau_{32}$. To have this requirement met, either the third laser has to be equally distant from the other two or artificial mechanisms have to be used to adjusting delay magnitudes, such as electronic buffering of the signals. Either way, the setup becomes more demanding and complicated.

As a result of the injection-locking effect, stable isochronal synchronization can be obtained between IRO1 and IRO2 due to the drive signal from IRO3 and communication can be established through the mutual coupling established between IRO1 and IRO2. The authors in Ref. [18] indicate that synchronization is maintained as the messages $m_1(t - \tau_{12})$ and $m_2(t - \tau_{21})$ are injected into the transmitting ring cavity, which are delayed by the propagation times τ_{12} and τ_{21}, respectively. According to the study, this procedure guarantees that the disturbance caused by the message injection will be symmetric and synchronization will be preserved. Another important factor to be considered is the relation among the coupling strengths. While K_{31} and K_{32} have to be chosen sufficiently large, since unidirectional drive induces isochronal synchronization between IRO1 and IRO2, due to injection-locking effect, the values of K_{12} and K_{21} should be chosen small enough not to disturb the synchronous state yet large enough to allow communication between IRO1 and IRO2.

Following a similar scheme, Jiang et al. [6] studied isochronal synchronization in the context of mutually coupled SLs subject to identical unidirectional injection. The results agree with previous ones in the sense that the presence of a sufficiently strong unidirectional coupling is needed for stable IS. In this sense, the strength of the unidirectional injection from SL3 was shown to be related to that of the mutual coupling: high-quality stable isochronal synchronization can be achieved as unidirectional injection from SL3 is sufficiently large as compared to mutual coupling between SL1 and SL2 [6]. This fact suggests that the coupling setups can be chosen from a large set of values, once they obbey certain symmetry and proportionality conditions, in opposition to self-feedback schemes which seem to offer smaller parameter ranges for stable synchronization, as shown from the results from Ref. [2].

2.3 Delayed Bidirectional Chain Coupling

More recent developments using three lasers show that isochronal synchronization is not only attainable under injection-locking effects. In fact, considering coupling setups featuring three lasers, all of them can settle down into isochronal synchronization, i.e., $x_1(t) = x_2(t) = x_3(t)$, as subject to bidirectional chain coupling. Consider the coupling scheme pictured in Fig. 4, based on the work of Illing et al. [5]. Note that the chain is coupled bidirectionally and all couplings are time-delayed. Considering this setup, a main requirement for IS to establish among the three oscillators is that the middle laser must be subject to self-feedback. Interestingly enough, for the coupling configuration presented in Fig. 4, isochronal synchronization is not observed between the center and outer oscillators in the absence of self-feedback, i.e., $K_{22} = 0$, despite of the symmetric nature of the coupling configuration. Such phenomenon had been studied earlier by Landsman and Schwartz [9], who explained that a common solution exists for the outer subsystems due to internal symmetry. In this case, the Lyapunov exponents calculated with respect to directions which are transversal to the synchronization manifold $s(t)$ are all negative, indicating that synchronization of the outer subsystems would emerge, i.e., $x_1(t) = x_3(t) = s(t)$. This coupling configuration is shown in Fig. 5, while the isochronal synchronization of the dynamics of outer lasers is pictured in Fig. 6.

Consider the coupling setup shown in Fig. 4. As the oscillators' equations are written as

$$\dot{x}_1(t) = Ax_1(t) + g(x_1(t)) + K_{21}x_2(t - \tau_{21}) \tag{20}$$

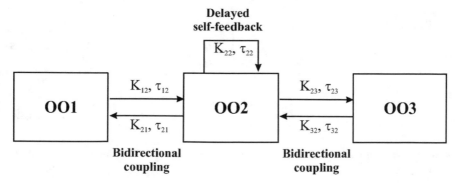

Fig. 4 Chain of optoelectronic oscillators (OO) bidirectionally coupled with time-delays τ_{12}, τ_{21}, τ_{23}, τ_{32} and self-feedback strength τ_{22}, with respective coupling strengths K_{12}, K_{21}, K_{23}, K_{32} and K_{22}: for zero self-coupling ($K_{22} = 0$), balanced bidirectional coupling ($K_{21} = K_{23}$, $K_{12} = K_{32}$) and equal coupling delays ($\tau_{12} = \tau_{21} = \tau_{23} = \tau_{32}$), the conditions are created for the emergence of isochronal synchronization between OO1 and OO3; as the self-feedback strength is raised above a threshold, isochronal synchronization among the three oscillators is possible [5]

$$\dot{x}_2(t) = Ax_2(t) + g(x_2(t)) + K_{12}x_1(t - \tau_{12}) + K_{22}x_2(t - \tau_{22}) + K_{32}x_3(t - \tau_{32})$$
(21)

$$\dot{x}_3(t) = Ax_3(t) + g(x_3(t)) + K_{23}x_2(t - \tau_{23})$$
(22)

As the synchronization between the outer systems is aimed, the error system is defined as $e = x_1 - x_3$. As

$$x(t) = \begin{bmatrix} x_1(t) \\ x_3(t) \end{bmatrix}$$
(23)

and

$$M = \begin{bmatrix} 1 & -1 \end{bmatrix} \otimes I_n$$
(24)

the synchronization manifold can be defined as in Eq. 5. Further, the error system can be written as

$$\dot{x}_1(t) - \dot{x}_3(t) = Ax_1(t) - Ax_3(t) + g(x_1(t)) - g(x_3(t)) + K_{21}x_2(t - \tau_{21}) + K_{23}x_2(t - \tau_{23})$$
(25)

Suppose $x_1(t) = x_3(t) = s(t)$, and the existence of the synchronization manifold is conditioned by

$$K_{21}s(t - \tau_{21}) = K_{23}s(s - \tau_{23})$$
(26)

which holds for $K_{21} = K_{23}$ and $\tau_{21} = \tau_{23}$.

One step ahead, as the synchronization of all systems is aimed, the error equations can be defined as

$$e_1 = x_1 - x_2$$
$$e_2 = x_1 - x_3$$
(27)

Define

$$x(t) = \begin{bmatrix} x_1(t) \\ x_2(t) \\ x_3(t) \end{bmatrix}$$
(28)

and

$$M = \begin{bmatrix} 1 & -1 & 0 \\ 1 & 0 & -1 \end{bmatrix} \otimes I_n$$
(29)

and the synchronization manifold can be defined as in Eq. 5. Further, the error system is given by

$$\dot{x}_1(t) - \dot{x}_2(t) = Ax_1(t) - Ax_2(t) + g(x_1(t)) - g(x_2(t)) + K_{21}x_2(t - \tau_{21})$$
$$- K_{22}x_2(t - \tau_{22}) - K_{12}x_1(t - \tau_{12}) - K_{32}x_3(t - \tau_{32})$$
$$\dot{x}_1(t) - \dot{x}_3(t) = Ax_1(t) - Ax_3(t) + g(x_1(t)) - g(x_3(t)) + K_{21}x_2(t - \tau_{21})$$
$$- K_{23}x_2(t - \tau_{23})$$
(30)

Suppose $x_1(t) = x_2(t) = x_3(t) = s(t)$, and then the existence of the synchronization manifold requires that

$$K_{21}s(t - \tau_{21}) = K_{22}s(t - \tau_{22}) + K_{12}s(t - \tau_{12}) + K_{32}s(t - \tau_{32}) \qquad (31)$$

and

$$K_{21}x_2(t - \tau_{21}) = K_{23}x_2(t - \tau_{23}) \qquad (32)$$

From Eq. 32, one obtains

$$K_{21} = K_{23}$$
$$\tau_{21} = \tau_{23} \qquad (33)$$

while Eq. 31 admits one of the following

$$(34) \quad \begin{cases} K_{21} = K_{22}, \ K_{12} = -K_{32} \\ \tau_{21} = \tau_{22}, \ \tau_{12} = \tau_{32} \\ K_{21} = K_{12}, \ K_{22} = -K_{32} \\ \tau_{21} = \tau_{12}, \ \tau_{22} = \tau_{32} \\ K_{21} = K_{32}, \ K_{22} = -K_{12} \\ \tau_{21} = \tau_{32}, \ \tau_{22} = \tau_{12} \\ K_{21} = K_{22} + K_{12} + K_{32} \\ \tau_{21} = \tau_{22} = \tau_{12} = \tau_{32} \end{cases}$$

Now consider the coupling setup presented in Fig. 5, whose equations can be supposed to have the form

$$\dot{x}_1(t) = Ax_1(t) + g\,(x_1(t)) + K_{21}x_2(t - \tau_{21}) \qquad (35)$$

$$\dot{x}_2(t) = Bx_2(t) + h\,(x_2(t)) + K_{12}x_1(t - \tau_{12}) + K_{32}x_3(t - \tau_{32}) \qquad (36)$$

$$\dot{x}_3(t) = Ax_3(t) + g\,(x_3(t)) + K_{23}x_2(t - \tau_{23}) \qquad (37)$$

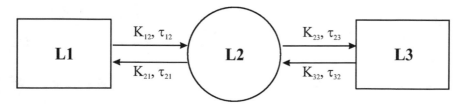

Fig. 5 Chain coupling of three lasers, as proposed in [9]: the outer lasers are coupled to the central laser by means of bidirectional coupling. For isochronal synchronization, coupling delays have to be symmetric, i.e., $\tau_{12} = \tau_{21} = \tau_{23} = \tau_{32}$. As sufficient coupling strength is provided, the dynamics of the identical lasers SL1 and SL3 can relax into isochronal synchrony. Within this framework, the central oscillator (SL2) can be different from the other two [16]

As the synchronization between systems 1 and 3 is aimed, the error equation can be defined as $e(t) = x_1(t) - x_3(t)$. Define

$$x(t) = \begin{bmatrix} x_1(t) \\ x_3(t) \end{bmatrix} \tag{38}$$

and

$$M = \begin{bmatrix} 1 & -1 \end{bmatrix} \otimes I_n \tag{39}$$

then the synchronization manifold can be defined as in Eq. 5. Further, the error system can be written as

$$\dot{x}_1(t) - \dot{x}_3(t) = Ax_1(t) - Ax_3(t) + g\left(x_1(t)\right) - g\left(x_3(t)\right) + K_{21}x_2(t - \tau_{21}) \\ - K_{23}x_2(t - \tau_{23}) \tag{40}$$

and, admitting $x_1(t) = x_3(t) = s(t)$, the existence of the synchronization manifold requires that

$$K_{21}x_2(t - \tau_{21}) = K_{23}x_2(t - \tau_{23}) \tag{41}$$

or simply

$$\begin{cases} K_{21} = K_{23} \\ \tau_{21} = \tau_{23} \end{cases} \tag{42}$$

For the same coupling configuration, suppose that the synchronization among all three systems is aimed. As such, the error equations can be defined as

$$\begin{aligned} e_1(t) &= x_1(t) - x_2(t) \\ e_2(t) &= x_1(t) - x_3(t) \end{aligned} \tag{43}$$

Further, as one defines

$$x(t) = \begin{bmatrix} x_1(t) \\ x_2(t) \\ x_3(t) \end{bmatrix} \tag{44}$$

and

$$M = \begin{bmatrix} 1 & -1 & 0 \\ 1 & 0 & -1 \end{bmatrix} \otimes I_n \tag{45}$$

the synchronization manifold can be defined as in Eq. 5. It follows that the error system can be written as

$$\begin{aligned} \dot{x}_1(t) - \dot{x}_2(t) &= Ax_1(t) - Bx_2(t) + g\left(x_1(t)\right) - h\left(x_2(t)\right) + K_{21}x_2(t - \tau_{21}) \\ & \quad - K_{12}x_1(t - \tau_{12}) - K_{32}x_3(t - \tau_{32}) \\ \dot{x}_1(t) - \dot{x}_3(t) &= Ax_1(t) - Ax_3(t) + g\left(x_1(t)\right) - g\left(x_3(t)\right) + K_{21}x_2(t - \tau_{21}) \\ & \quad - K_{23}x_2(t - \tau_{23}) \end{aligned} \tag{46}$$

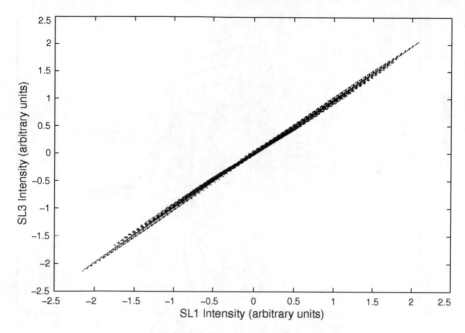

Fig. 6 Isochronal synchronization between outer lasers SL1 and SL3 in a three-laser chain: even in the absence of direct coupling between the oscillators, they can be driven to isochronal synchronization, by means of the interplay between the central and outer lasers

Supondo $x_1(t) = x_2(t) = x_3(t) = s(t)$, the existence of the synchronization manifold requires that

$$(A - B)s(t) + g\,(s(t)) - h\,(s(t)) + K_{21}s(t - \tau_{21}) - K_{12}s(t - \tau_{12}) - K_{32}s(t - \tau_{32}) = 0 \tag{47}$$

$$K_{21}s(t - \tau_{21}) = K_{23}s(t - \tau_{23}) \tag{48}$$

From Eq. 48, it is required that

$$\begin{cases} K_{21} = K_{23} \\ \tau_{21} = \tau_{23} \end{cases} \tag{49}$$

while Eq. 47 requires that

$$\begin{cases} A = B \\ g(.) = h(.) \\ K_{21} = K_{12} + K_{32} \\ \tau_{21} = \tau_{12} = \tau_{32} \end{cases} \tag{50}$$

from where it is concluded that the oscillators must be identical (Fig. 7).

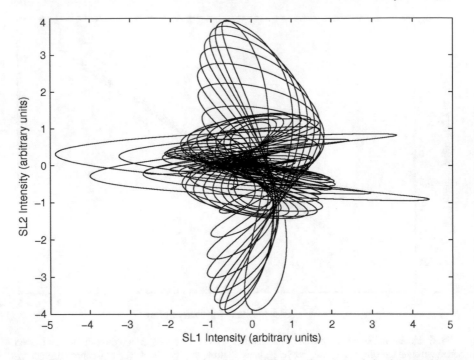

Fig. 7 Unsynchronized dynamical behavior of SL1 and SL2 in a three-laser chain: the central laser in the chain does not relax into isochronal synchronization as it is not subject to self-coupling, i.e., $x_1(t) \neq x_2(t)$ as $K_{22} = 0$

As further contribution, the authors [5] also showed that an analysis of such simple coupling configuration can lead to more general and profound comprehension of isochronal synchronization in networks of delay-coupled oscillators in general. As such, in a sense, simple configurations featuring three delay-coupled lasers can unveil mechanisms leading to synchronization in large networks. This is an important direction of research, since the understanding of stability of isochronal synchronization in networks may allow the extension of current chaos-based communication techniques to such realm [3, 4].

As observed in the results of aforementioned works, isochronal synchronization is highly dependent on symmetry of the coupling, as those in Figs. 1, 3, 4 and 5. On the one hand, its emergence requires that time-delays and the sum of coupling inputs are nearly identical. This may become a drawback in practice: while the coupling strength is a matter of design, time delays do not depend only upon design choices. This somewhat restricts the direct application of isochronal synchronization for communication in many forms of coupling setups featuring asymmetric characteristics, including most networks arising in real situations. In some cases, time delays can be electronically compensated and balanced by buffering signals, in order to adjusting time-delays for the design of communication systems for specific purposes, as in the case when spatial asymmetry is present. However, it is undeniable that the stringent requirement for symmetry adds an extra load of complexity to the practical realization of communication.

2.4 Mutual Delayed Coupling and Direct Self-feedback

Regarding two mutually coupled systems subject to direct self-feedback, according to the setup presented in Fig. 8, stable isochronal synchronization can be obtained and, beyond, it can be proven stable by means of analytical results [2, 11]. Note that direct self-feedback makes the practical implementation of the coupling setup easier, since direct self-feedback is easier to implement than delayed self-feedback. In practice, direct self-feedback offers a means for stable IS upon a coupling configuration that can be considered the simplest one. Note that the condition on the symmetry of delay magnitudes is easily respected in this case, and isochronal synchronization can emerge on the basis of the addition of mutual delayed coupling. Figures 9 and 10 show the dynamics of Lang-Kobayashi equations and the dynamics of the synchronization, respectively, as systems get in synchrony by mutual delayed feedback and direct self-feedback. On its turn, Fig. 11 presents the classic signature of synchronization, as state variables of the two systems are plot one against the other.

Consider the systems equations as given by

$$\dot{x}_1(t) = Ax_1(t) + g\left(x_1(t)\right) + K_{11}x_1(t - \tau_{11}) + K_{21}x_2(t - \tau_{21}) \tag{51}$$

$$\dot{x}_2(t) = Ax_2(t) + g\left(x_2(t)\right) + K_{22}x_2(t - \tau_{22}) + K_{12}x_1(t - \tau_{12}) \tag{52}$$

Note that, for direct self-feedback, $\tau_{11} = \tau_{22} = 0$. Define

$$x(t) = \begin{bmatrix} x_1(t) \\ x_2(t) \end{bmatrix} \tag{53}$$

$$M = \begin{bmatrix} 1 & -1 \end{bmatrix} \otimes I_n \tag{54}$$

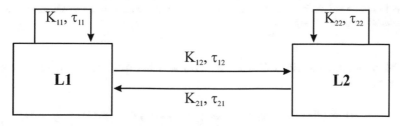

Fig. 8 Two mutually delay-coupled systems featuring direct self-feedback ($\tau_{11} = \tau_{22} = 0$): although simpler than most of the coupling setups studied in the literature, this coupling setup may severely restrict the magnitude of coupling delays

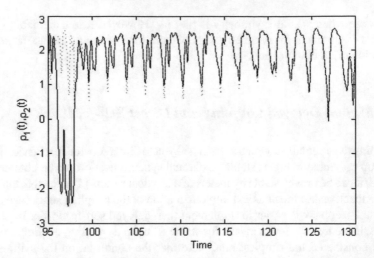

Fig. 9 Isochronal synchronization between two mutually coupled Lang-Kobayashi equations with direct self-feedback: coupling delay is taken as $\tau_{12} = \tau_{21} = 10^{-2}$

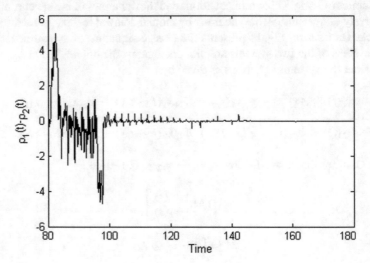

Fig. 10 Isochronal synchronization error for mutually delay-coupled Lang-Kobayashi equations subject to self-coupling: the error stabilizes at the origin of the error system, indicating that synchronous behavior established

and the synchronization manifold can be written as in Eq. 5. Further, define the error system as

$$
\begin{aligned}
\dot{x}_1(t) - \dot{x}_2(t) = {} & (A + K_{11})\, x_1(t) - (A + K_{22})\, x_2(t) + g\,(x_1(t)) - g\,(x_2(t)) \\
& + K_{21} x_2(t - \tau_{21}) - K_{12} x_1(t - \tau_{12})
\end{aligned}
\tag{55}
$$

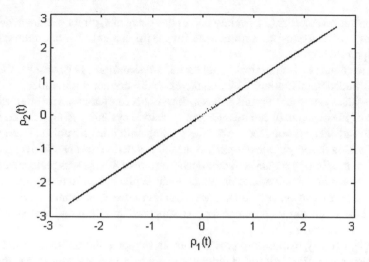

Fig. 11 Identity of trajectories of the mutually delay-coupled Lang-Kobayashi equations subject to direct self-feedback

and suppose $Mx(t) = 0$ holds, which implies $x_1(t) = x_2(t) = s(t)$, such that Eq. 55 can be rewritten as

$$\dot{s}(t) - \dot{s}(t) = (A + K_{11}) s(t) - (A + K_{22}) s(t) + g(s(t)) - g(s(t))$$
$$+ K_{21} s(t - \tau_{21}) - K_{12} s(t - \tau_{12})$$
(56)

It follows that the existence of S is conditioned to

$$(K_{11} - K_{22}) s(t) + K_{21} s(t - \tau_{21}) - K_{12} s(t - \tau_{12}) = 0$$
(57)

Regarding the system
$$\dot{s}(t) = As(t) + g(s(t))$$
(58)

as chaotic, it follows that Eq. 57 holds under $K_{11} = K_{22}$, $K_{21} = K_{12}$ and $\tau_{21} = \tau_{12}$.

Under frameworks based on direct self-couplings, the requirement that time delay of self-couplings are adjusted for symmetric operation can be relaxed, since this condition ceases to exist in this case; on the other hand, this framework seems to impose an upper bound on the allowable value of delay in the mutual coupling such that the chaotic behavior of the coupled systems and synchronization stability are preserved. This is undesirable under the viewpoint of communication, since upper bounded coupling delays mean that communication cannot be established if the distance between oscillators is large. The root of the problem is that isochronal synchronization becomes unfeasible for large delays under this framework. In addition, the parameter ranges for stability seemingly becomes narrower under direct feedback setups and, more importantly, time-delays become more influential upon synchronization stability [2]. Mainly due to these reasons, direct self-feedback does not seem to offer a consistent framework for the implementation of communication based upon IS.

One step ahead of numerical simulations, Wagemakers [17] conceived the first experimental demonstration of simultaneous bidirectional communication. Using Mackey-Glass electronic circuits as basis, the authors implemented full-duplex transmission of messages between delay-coupled chaotic systems, using delayed mutual coupling and delayed self-feedback. After dealing with intrinsic noise of the systems and evaluating the amplitude of the message to be added in the chaotic dynamics, the transmission of binary sequences at the encoding rate of 80 *bps* was performed at both ends of the communication line, simultaneously. While not particularly attractive for its bit rate performance, such implementation promoted experimental evidence that IS can indeed sustain communication, as supposed in computational simulations using laser equations [15, 18].

Needless to say, while they provide simple implementation, chaotic electronic circuits are not particularly suited for applications in communication due to their low encoding rate capacity. In this context, lasers have superior encoding performance

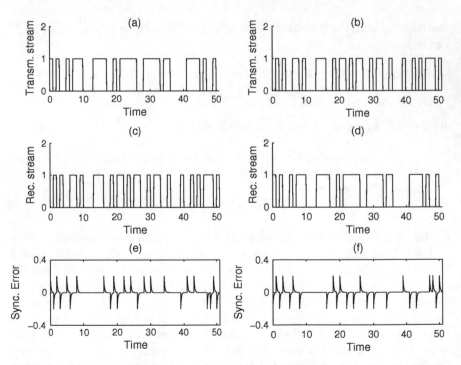

Fig. 12 Numerical realization of bidirectional simultaneous transmission of binary messages, using the methodology proposed in [17]: **a** binary message transmitted, **b** received by system 1 and **c** synchronization error, as measured in system 1; **d** binary message transmitted and **e** received by system 2 and (**c**) synchronization error, as measured in system 2

while they also benefit from low noise levels that are a desirable feature of optical systems in general. Thus, due to their operation at time scales of the order of nano to picoseconds (10^{-9} to 10^{-12} s), lasers can be subject to much higher injection rates while they also benefit from low noise level of optical fiber as compared to that of copper-based systems [13]. As such, the next step towards the development of simultaneous bidirectional transmission using isochronal synchronization seems to be its experimental implementation using lasers (Fig. 12).

In a broad context, beyond the realm of lasers or electronic circuits, isochronal synchronization was analytically shown to be achievable in pairs of delay-coupled chaotic oscillators in general, indicating that its occurrence, as physical phenomenon, is not restricted to a particular chaotic oscillator [2]. Further, the phenomenon was shown to emerge even in network configurations, in prospective studies aiming its application for communication in flying formation of satellites [3, 4]. Seemingly, the symmetry of oscillator couplings, both regarding its strength and delay magnitude, seems to be a stringent requirement in the context of networks as well. Moreover, as a synchronous isochronal solution does not exist for every network topology, particularly as one considers coupling delays and direct self-feedback, as shown in [11], it seems that the development of network communication schemes based on isochronal synchronization is summarily restricted to specific topologies that are not commonly found in real networks. These conditions seem to be mainly related to network symmetry conditions [3, 4, 11] (Figs. 13, 14 and 15).

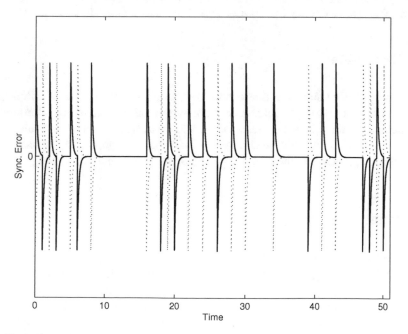

Fig. 13 Anti-synchronization between the error variables from each of the systems against time: the error dynamics in both ends of the communication channel remain synchronized as random binary streams are input simultaneously in the dynamics of both chaotic oscillators

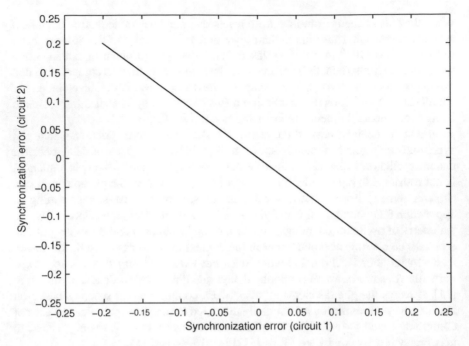

Fig. 14 Anti-synchronization between error variables measures in each of the systems, against each other: persistent synchronous behavior shows that simultaneous bidirectional communication can be exploited in mutually delay-coupled chaotic oscillators, as experimentally demonstrated in [17]

3 Discussion

This study showed that the conditions for the existence of the synchronization manifold is associated to the establishment of homogeneous coupling inputs throughout the network nodes whose dynamics belongs to the synchronization manifold. Once such condition is satisfied, a synchronous solution can emerge, depending on specific circumstances related to the coupling strength, delay and network topology.

4 Final Remarks

The present study evaluated the conditions for the existence of a synchronization manifold in coupling setups recently explored for bidirectional simultaneous communication using isochronal chaos synchronization. As main result, strict mathematical relations among the coupling setup parameters were established, leading to one general conclusion concerning the existence of the synchronization manifold.

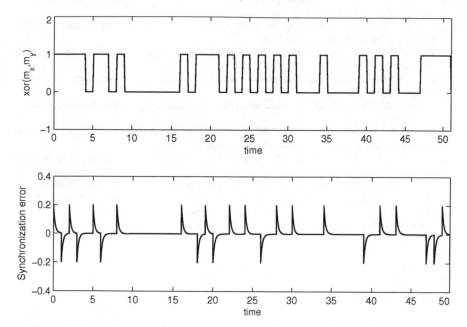

Fig. 15 XOR between the binary streams $m_x(t)$ and $m_y(t)$ and synchronization error $e_x(t)$: spikes in the synchronization error from zero indicate the encoding of different bits in each end of the communication channel

References

1. Deng T, Xia GQ, Cao LP, Chen JG, Lin XD, Wu ZM (2009) Bidirectional chaos synchronization and communication in semiconductor lasers with optoelectronic feedback. Opt Commun 282:2243–2249
2. Grzybowski JMV, Macau EEN, Yoneyama T (2011) Isochronal synchronization of time delay and delay-coupled chaotic systems. J Phys A: Math Theor 44:175103
3. Grzybowski JMV, Rafikov M, Macau EEN (2009) Chaotic communication in a satellite formation flying—the synchronization issue in a scenario with transmission delay. Acta Astronaut 14:2793–2806
4. Grzybowski JMV, Rafikov M, Macau EEN (2010) Synchronization analysis for chaotic communication on a satellite formation flying. Acta Astronautica 67:881–891
5. Illing L, Panda CD, Shareshian L (2011) Isochronal chaos synchronization of delay-coupled optoelectronic oscillators. Phys Rev E 84:016213
6. Jiang N, Pan W, Yan L, Luo B, Zhang W, Xiang S, Yang L, Zheng D (2010) Chaos synchronization and communication in mutually coupled semiconductor lasers driven by a third laser. J Lightwave Technol 28:1978–1985
7. Kanter I, Kopelowitz E, Kinzel W (2008) Public channel cryptography: chaos synchronization and Hilbert's tenth problem. Phys Rev Lett 101:084102
8. Klein E, Gross N, Rosenbluh M, Kinzel W, Khaytovich L, Kanter I (2006) Stable isochronal synchronization of mutually coupled chaotic lasers. Phys Rev E 73:066214
9. Landsman A, Schwartz IB (2007) Complete chaotic synchronization in mutually coupled time-delay systems. Phys Rev E 75:026201
10. Oguchi T, Nijmeijer H, Yamamoto T (2007) Synchronization of coupled nonlinear systems with time delay. In: Proceedings of the European control conference, pp 3056–3061

11. Oguchi T, Nijmeijer H, Yamamoto T (2008) Synchronization in networks of chaotic systems with delay-coupling. Chaos 18:037108
12. Oguchi T, Nijmeijer H, Yamamoto T, Kniknie T (2008) Synchronization of four identical nonlinear systems with time-delay, pp 12153–12158
13. Prasad KV (2004) Principles of digital communication systems and computer networks. Charles River Media
14. Schwartz IB, Shaw LB (2006) Isochronal synchronization of delay-coupled systems. Phys Rev E 73:046207
15. Vicente R, Mirasso CR, Fischer I (2007) Simultaneous bidirectional message transmission in a chaos-based communication scheme. Opt Lett 32:403–405
16. Wagemakers A, Buldu JM, Sanjuan MAF (2007) Isochronous synchronization in mutually coupled chaotic circuits. Chaos 17:023128
17. Wagemakers A, Buldu JM, Sanjuan MAF (2008) Experimental demonstration of bidirectional chaotic communication by means of isochronal synchronization. Europhys Lett 81:40005
18. Zhou BB, Roy R (2007) Isochronal synchrony and bidirectional communication with delay-coupled nonlinear oscillators. Phys Rev E 75:026205

A Sequential Order of Periodic Motions in a 1-Dimensional, Time-Delay, Dynamical System

Siyuan Xing and Albert C. J. Luo

Abstract In this chapter, periodic motions in a 1-D, time-delayed, nonlinear dynamical system are explored by a semi-analytical method. The stability and bifurcations of periodic motions are determined by eigenvalue analysis. A global sequence of periodic motions $P_{1(S)} \lhd P_{1(A)} \lhd P_{3(S)} \lhd P_{2(A)} \lhd \ldots \lhd P_{m(A)} \lhd P_{2m-1(S)} \lhd \ldots (m = 0, 1, 2, \ldots)$ is discovered. Numerical simulations of periodic motions are performed from analytical predictions. From finite Fourier series analysis, harmonic amplitude and phases for periodic motions are presented. This chapter is dedicated to Valentin Afraimovich for a friendship between Valentin Afraimovich and Albert Luo during the past 20 years. The results are from the useful and valuable discussions between Valentin and Albert.

1 Introduction

In recent years, time-delay, nonlinear dynamical systems are of great interest in physiology, optics and engineering. Mackey and Glass [1] numerically investigated periodic motions and aperiodic motions in a 1-D, time-delayed nonlinear dynamical to explain physiological diseases. Namajūnas [2] experimentally observed chaotic motions in such a 1-D nonlinear system through an analog circuit with a tuning delay component. Ikeda et al. [3, 4] introduced a time-delay model of an optical bi-stable resonator and observed multiple stable periodic motions in such a system through numerical simulation. The possibility of a memory device for information storage by such periodic motions were discussed. Yang et al. [5] studied the vibration resonance in a Duffing system with a generalized delay feedback represented in a fractional-order differential form. Jeevarathinam et al. [6] further investigated the

S. Xing
Department of Mechanical Engineering, California Polytechnic State University, San Luis Obispo CA93401, USA

A. C. J. Luo (✉)
Department of Mechanical and Mechatronics Engineering, Southern Illinois University Edwardsville, Edwardsville IL62026-1805, USA
e-mail: aluo@siue.edu

© Higher Education Press 2021
D. Volchenkov (ed.), *Nonlinear Dynamics, Chaos, and Complexity*,
Nonlinear Physical Science, https://doi.org/10.1007/978-981-15-9034-4_7

Duffing systems with a gamma distributed time-delayed feedback and an integrative time-delayed feedback, respectively. Time-delay systems have been also studied for synchronization. For instance, Krishnaveni and Sathiyanathan [7] investigated the synchronization of couple map lattice using delayed variable feedback. Akhmet [8] studied the synchronization of the cardiac pacemaker model with delayed pulse-coupling.

To analytically predict periodic motions in time-delayed, nonlinear dynamical systems, one employed the perturbation methods (e.g., [9, 10]) and harmonic balance method (e.g., [11, 12]). Such methods are adequate only for simple periodic motions, but insufficient for complex periodic motions. Luo [13] introduced the generalized harmonic balance method, which could be used for analytical solutions of periodic motions and even for chaos. Luo and Huang [14] applied such a method for periodic motions in a Duffing oscillator. Luo [15] extended the generalized harmonic balanced method for time-delayed, nonlinear systems. Luo and Jin [16] used such a method for bifurcation trees of period-m motion in a quadratic nonlinear oscillator with time-delay.

For the generalized harmonic balance method, it is difficult to determine periodic motions in dynamical systems with non-polynomial nonlinearity. Hence, Luo [17] developed a semi-analytical method to determine periodic motions in nonlinear systems with/without delay. Such an approach formulated discrete implicit mappings for periodic motions with a controllable global error. Luo and Guo [18] applied such a method for periodic motions in a nonlinear Duffing oscillator, and the corresponding results were identical to those results from the generalized harmonic balance method. Luo and Xing [19] employed such a semi-analytical method for bifurcation trees of period-1 motions to chaos in a time-delayed, hardening Duffing oscillator. Multiple complex bifurcation trees of period-1 motions to chaos were obtained. Luo and Xing [20] studied periodic motions in a twin-well, Duffing oscillator with time-delay displacement feedback and discussed the possibility of infinite bifurcation trees in such a nonlinear oscillator. Luo and Xing [21] also investigated the time-delay effects on periodic motions in a time-delayed, Duffing oscillator. Luo and Xu [22] discovered the series order of periodic motions in a two-degree-of-freedom van der Pol-Duffing oscillator.

To better understand dynamic behaviors of 1-dimensional, time-delayed, nonlinear dynamical systems, consider the following system as

$$\dot{x} = \alpha_1 x - \alpha_2 \sin x^\tau - \beta x^3 + Q_0 \cos \Omega t, \tag{1}$$

where $x = x(t)$ and $x^\tau = x(t - \tau)$. α_1, α_2 and β are system coefficients. Ω and Q_0 are excitation frequency and excitation amplitude, respectively. Periodic motions in such a nonlinear system will be presented herein through the semi-analytical method.

2 Methodology

Theorem 1 [17] *Consider a time-delay nonlinear dynamical system.*

$$\dot{\mathbf{x}} = \mathbf{f}(\mathbf{x}, \mathbf{x}^\tau, t, \mathbf{p}) \in \mathscr{R}^n,$$

$$\text{with } \mathbf{x}(t_0) = \mathbf{x}_0, \ \mathbf{x}(t) = \mathbf{\Phi}(\mathbf{x}_0, t - t_0, \mathbf{p}) \text{ for } t \in [t_0 - \tau, \infty). \tag{2}$$

If such a time-delay dynamical system has a period-m flow $\mathbf{x}^{(m)}(t)$ with finite norm $\|\mathbf{x}^{(m)}\|$ and period mT $(T = 2\pi/\Omega)$, there is a set of discrete time $t_k(k = 0, 1, \ldots, mN)$ with $(N \to \infty)$ during m-periods (mT), and the corresponding solution $\mathbf{x}^{(m)}(t_k)$ and vector fields $\mathbf{f}(\mathbf{x}^{(m)}(t_k), \mathbf{x}^{\tau(m)}(t_k), t_k, \mathbf{p})$ are exact. Suppose discrete nodes $\mathbf{x}_k^{(m)}$ and $\mathbf{x}_k^{\tau(m)}$ are on the approximate solution of the periodic flow under $\|\mathbf{x}^{(m)}(t_k) - \mathbf{x}_k^{(m)}\| \le \varepsilon_k$ and $\|\mathbf{x}^{\tau(m)}(t_k) - \mathbf{x}_k^{\tau(m)}\| \le \varepsilon_k^\tau$ with small $\varepsilon_k, \varepsilon_k^\tau \ge 0$ and

$$\|\mathbf{f}(\mathbf{x}^{(m)}(t_k), \mathbf{x}^{\tau(m)}(t_k), t_k, \mathbf{p}) - \mathbf{f}(\mathbf{x}_k^{(m)}, \mathbf{x}_k^{\tau(m)} t_k, \mathbf{p})\| \le \delta_k \tag{3}$$

with a small $\delta_k \ge 0$. During a time interval $t \in [t_{k-1}, t_k]$, there is a mapping $P_k : (\mathbf{x}_{k-1}^{(m)}, \mathbf{x}_{k-1}^{\tau(m)}) \to (\mathbf{x}_k^{(m)}, \mathbf{x}_k^{\tau(m)})$ $(k = 1, 2, \ldots, mN)$ as

$$(\mathbf{x}_k^{(m)}, \mathbf{x}_k^{\tau(m)}) = P_k(\mathbf{x}_{k-1}^{(m)}, \mathbf{x}_{k-1}^{\tau(m)}) \text{ with } \mathbf{g}_k(\mathbf{x}_{k-1}^{(m)}, \mathbf{x}_k^{(m)}; \mathbf{x}_{k-1}^{\tau(m)}, \mathbf{x}_k^{\tau(m)}, \mathbf{p}) = \mathbf{0},$$

$$\mathbf{x}_j^{\tau(m)} = \mathbf{h}_j(\mathbf{x}_{r_j-1}^{(m)}, \mathbf{x}_{r_j}^{(m)}, \theta_{r_j}), \ j = k, k-1; r_j = j - l_j, \ k = 1, 2, \ldots, mN;$$

$$(e.g., \mathbf{x}_r^{\tau(m)} = \mathbf{x}_{s_r}^{(m)} + \theta_r(\mathbf{x}_{r_r-1}^{(m)} - \mathbf{x}_{r_r}^{(m)}), \theta_r = \frac{1}{h_{r_j}}[\tau - \sum_{i=1}^{l_{r_j}} h_{r_j+i}]). \tag{4}$$

where $\mathbf{g}_k$ is an implicit vector function and $\mathbf{h}_j$ is an interpolation vector function. Consider a mapping structure as

$$P = P_{mN} \circ P_{mN-1} \circ \ldots \circ P_2 \circ P_1 : \mathbf{x}_0^{(m)} \to \mathbf{x}_{mN}^{(m)};$$

$$\text{with } P_k : (\mathbf{x}_{k-1}^{(m)}, \mathbf{x}_{k-1}^{\tau(m)}) \to (\mathbf{x}_k^{(m)}, \mathbf{x}_k^{\tau(m)}) \ (k = 1, 2, \ldots, mN). \tag{5}$$

For $\mathbf{x}_{mN}^{(m)} = P(\mathbf{x}_0^{(m)}, \mathbf{x}_0^{\tau(m)})$, if there is a set of points $(\mathbf{x}_k^{(m)*}, \mathbf{x}_k^{\tau(m)*})(k = 0, 1, \ldots, mN)$ computed by

$$\left.\begin{array}{l} \mathbf{g}_k(\mathbf{x}_{k-1}^{(m)*}, \mathbf{x}_k^{(m)*}; \mathbf{x}_{k-1}^{\tau(m)*}, \mathbf{x}_k^{\tau(m)*}, \mathbf{p}) = \mathbf{0}, \\ \mathbf{x}_j^{\tau(m)*} = \mathbf{h}_j(\mathbf{x}_{r_j-1}^{(m)*}, \mathbf{x}_{r_j}^{(m)*}, \theta_{r_j}), \ j = k, k-1 \end{array}\right\} (k = 1, 2, \ldots, mN)$$

$$\mathbf{x}_{r_j-1}^{(m)*} = \mathbf{x}_{\text{mod}(r_j-1+mN, mN)}^{(m)*}, \mathbf{x}_{r_j}^{(m)*} = \mathbf{x}_{\text{mod}(r_j+mN, mN)}^{(m)*};$$

$$\mathbf{x}_0^{(m)*} = \mathbf{x}_{mN}^{(m)*} \text{ and } \mathbf{x}_0^{\tau(m)*} = \mathbf{x}_{mN}^{\tau(m)*}, \tag{6}$$

then the points $\mathbf{x}_k^{(m)*}$ and $\mathbf{x}_k^{\tau(m)*}$ $(k = 0, 1, \ldots, mN)$ are the approximation of points $\mathbf{x}^{(m)}(t_k)$ and $\mathbf{x}^{\tau(m)}(t_k)$ of periodic solutions. In the neighborhoods of $\mathbf{x}_k^{(m)*}$ and $\mathbf{x}_k^{\tau(m)*}$, with $\mathbf{x}_k^{(m)} = \mathbf{x}_k^{(m)*} + \Delta\mathbf{x}_k^{(m)}$ and $\mathbf{x}_k^{\tau(m)} = \mathbf{x}_k^{\tau(m)*} + \Delta\mathbf{x}_k^{\tau(m)}$, the linearized equation is given by

$$\sum_{j=k-1}^{k} \frac{\partial \mathbf{g}_k}{\partial \mathbf{x}_j^{(m)}} \Delta\mathbf{x}_j^{(m)} + \frac{\partial \mathbf{g}_k}{\partial \mathbf{x}_j^{\tau(m)}} \left(\frac{\partial \mathbf{x}_j^{\tau(m)}}{\partial \mathbf{x}_{r_j}^{(m)}} \Delta\mathbf{x}_{r_j}^{(m)} + \frac{\partial \mathbf{x}_j^{\tau(m)}}{\partial \mathbf{x}_{r_j-1}^{(m)}} \Delta\mathbf{x}_{r_j-1}^{(m)} \right) = \mathbf{0}$$

with $r_j = j - l_j$, $j = k - 1, k$; $(k = 1, 2, \ldots, mN)$. $\qquad(7)$

The resultant Jacobian matrices of the periodic flow are

$$DP_{k(k-1)\ldots 1} = \left[\frac{\partial \mathbf{y}_k^{(m)}}{\partial \mathbf{y}_0^{(m)}} \right]_{(\mathbf{y}_0^{(m)*}, v, \mathbf{y}_k^{(m)*})} = \mathbf{A}_k \mathbf{A}_{k-1} \ldots \mathbf{A}_1 \ (k = 1, 2, \ldots, mN),$$

and $DP = DP_{mN(mN-1)\ldots 1} = \left[\frac{\partial \mathbf{y}_{mN}^{(m)}}{\partial \mathbf{y}_0^{(m)}} \right]_{(\mathbf{y}_0^{(m)*}, v, \mathbf{y}_{mN}^{(m)*})} = \mathbf{A}_{mN} \mathbf{A}_{mN-1} \ldots \mathbf{A}_1$ $\qquad(8)$

where

$$\Delta\mathbf{y}_k^{(m)} = \mathbf{A}_k^{(m)} \Delta\mathbf{y}_{k-1}^{(m)}, \ \mathbf{A}_k^{(m)} \equiv \left[\frac{\partial \mathbf{y}_k^{(m)}}{\partial \mathbf{y}_{k-1}^{(m)}} \right]_{(\mathbf{y}_{k-1}^{(m)*}, \mathbf{y}_k^{(m)*})} \qquad(9)$$

and

$$\mathbf{A}_k^{(m)} = \left[\begin{array}{cc} \mathbf{B}_k^{(m)} & (\mathbf{a}_{k(r_{k-1}-1)}^{(m)})_{n \times n} \\ \mathbf{I}_k^{(m)} & \mathbf{0}_k^{(m)} \end{array} \right]_{n(s+1) \times n(s+1)}, s = 1 + l_{k-1}$$

$$\mathbf{B}_k^{(m)} = [(\mathbf{a}_{k(k-1)}^{(m)})_{n \times n}, \mathbf{0}_{n \times n}, \ldots, (\mathbf{a}_{k(r_k-1)}^{(m)})_{n \times n}],$$

$$\mathbf{I}_k^{(m)} = \mathrm{diag}(\mathbf{I}_{n \times n}, \mathbf{I}_{n \times n}, \ldots, \mathbf{I}_{n \times n})_{ns \times ns},$$

$$\mathbf{0}_k^{(m)} = (\underbrace{\mathbf{0}_{n \times n}, \mathbf{0}_{n \times n} \ldots, \mathbf{0}_{n \times n}}_{s})^{\mathrm{T}}; \qquad(10)$$

$$\mathbf{y}_k^{(m)} = (\mathbf{x}_k^{(m)}, \mathbf{x}_{k-1}^{(m)}, \ldots, \mathbf{x}_{r_{k-1}}^{(m)})^{\mathrm{T}},$$

$$\mathbf{y}_{k-1}^{(m)} = (\mathbf{x}_{k-1}^{(m)}, \mathbf{x}_{k-2}^{(m)}, \ldots, \mathbf{x}_{r_{k-1}-1}^{(m)})^{\mathrm{T}},$$

$$\Delta\mathbf{y}_k^{(m)} = (\Delta\mathbf{x}_k^{(m)}, \Delta\mathbf{x}_{k-1}^{(m)}, \ldots, \Delta\mathbf{x}_{r_{k-1}}^{(m)})^{\mathrm{T}},$$

$$\Delta\mathbf{y}_{k-1}^{(m)} = (\Delta\mathbf{x}_{k-1}^{(m)}, \Delta\mathbf{x}_{k-2}^{(m)}, \ldots, \Delta\mathbf{x}_{r_{k-1}-1}^{(m)})^{\mathrm{T}}; \qquad(11)$$

$$\mathbf{a}_{kj}^{(m)} = -\left[\frac{\partial \mathbf{g}_k}{\partial \mathbf{x}_k^{(m)}}\right]^{-1} \frac{\partial \mathbf{g}_k}{\partial \mathbf{x}_j^{(m)}},$$

$$\mathbf{a}_{kr_j}^{(m)} = -\left[\frac{\partial \mathbf{g}_k}{\partial \mathbf{x}_k^{(m)}}\right]^{-1} \sum_{\alpha=j}^{j+1} \frac{\partial \mathbf{g}_k}{\partial \mathbf{x}_\alpha^{\tau(m)}} \frac{\partial \mathbf{x}_\alpha^{\tau(m)}}{\partial \mathbf{x}_{r_j}},$$

$$\mathbf{a}_{k(r_j-1)}^{(m)} = -\left[\frac{\partial \mathbf{g}_k}{\partial \mathbf{x}_k^{(m)}}\right]^{-1} \sum_{\alpha=j-1}^{j} \frac{\partial \mathbf{g}_k}{\partial \mathbf{x}_\alpha^{\tau(m)}} \frac{\partial \mathbf{x}_\alpha^{\tau(m)}}{\partial \mathbf{x}_{r_j-1}}$$

with $r_j = j - l_j, \ j = k - 1, k$.   (12)

The properties of discrete points $\mathbf{x}_k^{(m)}$ *$(k = 1, 2, \ldots, mN)$ can be estimated by the eigenvalues of* $DP_{k(k-1)\ldots1}$ *as*

$$\left|DP_{k(k-1)\nu1} - \overline{\lambda}\mathbf{I}_{n(s+1)\times n(s+1)}\right| = 0 \ (k = 1, 2, \ldots, mN).$$ (13)

The eigenvalues of DP for such periodic motion are determined by

$$|DP - \lambda \mathbf{I}_{n(s+1)\times n(s+1)}| = 0,$$ (14)

and the stability and bifurcation of the periodic flow can be classified by the eigenvalues of $DP(\mathbf{y}_0^*)$ *with*

$$([n_1^m, n_1^o] : [n_2^m, n_2^o] : [n_3, \kappa_3] : [n_4, \kappa_4]|n_5 : n_6 : [n_7, l, \kappa_7]).$$ (15)

(i) *If the magnitudes of all eigenvalues of DP are less than one (i.e.,$|\lambda_i| < 1$, $i = 1, 2, \ldots, n(s + 1)$), the approximate periodic solution is stable.*

(ii) *If at least the magnitude of one eigenvalue of DP is greater than one (i.e.,$|\lambda_i| > 1$, $i \in \{1, 2, \ldots, n(s + 1)\}$), the approximate periodic solution is unstable.*

(iii) *The boundaries between stable and unstable periodic flow with higher order singularity give bifurcation and stability conditions with higher order singularity.*

Proof: See Luo [17].   ∎

3 Implicit Discretization Scheme

Let $x_k = x(t_k)$ at $t_k = kh$ $(k = 0, 1, 2, \ldots)$ be the sampled nodes with uniform time-step on a periodic motion in the 1-dimensional, time-delay, nonlinear dynamical system. $x_k^\tau = x(t_k - \tau)$ is the corresponding delayed point of x_k. By the mid-point discretization scheme for the nonlinear dynamical system, an implicit map P_k from (x_{k-1}, x_{k-1}^τ) to (x_k, x_k^τ) is constructed as

$$P_k : (x_{k-1}, x_{k-1}^\tau) \to (x_k, x_k^\tau) \Rightarrow (x_k, x_k^\tau) = P_k(x_{k-1}, x_{k-1}^\tau). \qquad (16)$$

The mapping P_k is expressed as

$$x_k - x_{k-1} = h\{\tfrac{1}{2}\alpha_1(x_k + x_{k-1}) - \alpha_2 \sin[\tfrac{1}{2}(x_k^\tau + x_{k-1}^\tau)]$$
$$- \tfrac{1}{8}\beta(x_k + x_{k-1})^3 + Q_0 \cos \Omega(t + \tfrac{1}{2}h)\}. \qquad (17)$$

The time-delay node $x_j^\tau (j = k - 1, k)$ is interpolated by two adjacent points x_{r_j-1} and x_{r_j} as

$$x_j^\tau = h_j(x_{r_j-1}, x_{r_j}, \theta_{r_j}) \text{ for } r_j = j - l_j, \qquad (18)$$

where $l_k = \text{int}(\tau/h)$. Using a simple Lagrange interpolation, the time-delay discrete node $x_j^\tau = h_j(x_{r_j-1}, x_{r_j}, \theta_{r_j})$ $(j = k, k - 1)$ is represented by

$$x_k^\tau = x_{k-l_k-1} + (1 - \tfrac{1}{h}\tau + l_k)(x_{k-l_k} - x_{k-l_k-1}). \qquad (19)$$

Now the discretization of the delayed nonlinear dynamical system is performed.

4   Period-m Motions and Stability

In the 1-D, time-delay, nonlinear dynamical system, a period-m motion is described by a discrete mapping structure

$$P = \underbrace{P_{mN} \circ P_{mN-1} \circ \ldots \circ P_2 \circ P_1}_{mN-\text{actions}} : (x_0^{(m)}, x_0^{\tau(m)}) \to (x_{mN}^{(m)}, x_{mN}^{\tau(m)}),$$
$$(x_{mN}^{(m)}, x_{mN}^{\tau(m)}) = P(x_0^{(m)}, x_0^{\tau(m)}), \qquad (20)$$

with

$$P_k : (x_{k-1}^{(m)}, x_{k-1}^{\tau(m)}) \to (x_k^{(m)}, x_k^{\tau(m)})$$
$$(k = 1, 2, \ldots, mN). \qquad (21)$$

The discrete nodes $x_k^{(m)*}(k = 1, 2, \ldots, mN)$ on the period-m motion of the 1-D, time-delay nonlinear system are computed by

$$\left.\begin{array}{l} g_k(x_{k-1}^{(m)*}, x_k^{(m)*}, x_{k-1}^{\tau(m)*}, x_k^{\tau(m)*}, \mathbf{p}) = 0 \\ x_j^{\tau(m)*} = h_j(x_{r_j-1}^{(m)*}, x_{r_j}^{(m)*}, \theta_{r_j}), \quad j = k, k-1 \end{array}\right\} (k = 1, 2, \ldots, mN),$$

$$x_0^{(m)*} = x_{mN}^{(m)*} \text{ and } x_0^{\tau(m)*} = x_{mN}^{\tau(m)*}. \tag{22}$$

The corresponding algebraic equations of $g_k = 0$ $(k = 1, 2, \ldots, mN)$ in Eq. (22) are

$$g_k \equiv x_k^{(m)*} - x_{k-1}^{(m)*} - h\{\tfrac{1}{2}\alpha_1(x_k^{(m)*} + x_{k-1}^{(m)*}) - \alpha_2 \sin[\tfrac{1}{2}(x_k^{\tau(m)*} + x_{k-1}^{\tau(m)*})]$$
$$- \tfrac{1}{8}\beta(x_k^{(m)*} + x_{k-1}^{(m)*})^3 + Q_0 \cos \Omega (t + \tfrac{1}{2}h)\}$$
$$= 0. \tag{23}$$

The time-delay nodes $x_j^{\tau(m)*}$ $(j = k, k-1)$ in Eq. (22) are from Eq. (19), i.e.,

$$x_k^{\tau(m)*} = x_{k-l_k-1}^{(m)*} + (1 - \tfrac{1}{h}\tau + l_k)(x_{k-l_k}^{(m)*} - x_{k-l_k-1}^{(m)*}),$$
$$x_{k-1}^{\tau(m)*} = x_{k-1-l_{k-1}-1}^{(m)*} + (1 - \tfrac{1}{h}\tau + l_{k-1})(x_{k-1-l_{k-1}}^{(m)*} - x_{k-1-l_{k-1}-1}^{(m)*}). \tag{24}$$

From Eqs. (22) to (24), the approximate discrete nodes of period-m motions in the 1-D, time-delay, nonlinear dynamical system are determined. For the stability and bifurcation of the period-m motion, in vicinity of $x_k^{(m)*}$ and $x_k^{\tau(m)*}$, $x_k^{(m)} = x_k^{(m)*} + \Delta x_k^{(m)}$ and $x_k^{\tau(m)} = x_k^{\tau(m)*} + \Delta x_k^{\tau(m)}$. The linearized equations of implicit mappings are

$$\sum_{j=k-1}^{k} \frac{\partial g_k}{\partial x_j^{(m)}} \Delta x_j^{(m)} + \frac{\partial g_k}{\partial x_j^{\tau(m)}} (\frac{\partial x_j^{\tau(m)}}{\partial x_{r_j}^{(m)}} \Delta x_{r_j}^{(m)} + \frac{\partial x_j^{\tau(m)}}{\partial x_{r_j-1}^{(m)}} \Delta x_{r_j-1}^{(m)}) = 0,$$

with $r_j = j - l_j$, $j = k-1, k$; $(k = 1, 2, \ldots, mN)$. $\tag{25}$

Define

$$\mathbf{y}_k^{(m)} = (x_k^{(m)}, x_{k-1}^{(m)}, \ldots, x_{r_k-1}^{(m)})^{\mathrm{T}},$$
$$\mathbf{y}_{k-1}^{(m)} = (x_{k-1}^{(m)}, x_{k-2}^{(m)}, \ldots, x_{r_{k-1}-1}^{(m)})^{\mathrm{T}},$$
$$\Delta\mathbf{y}_k^{(m)} = (\Delta x_k^{(m)}, \Delta x_{k-1}^{(m)}, \ldots, \Delta x_{r_k-1}^{(m)})^{\mathrm{T}},$$
$$\Delta\mathbf{y}_{k-1}^{(m)} = (\Delta x_{k-1}^{(m)}, \Delta x_{k-1}^{(m)}, \ldots, \Delta x_{r_{k-1}-1}^{(m)})^{\mathrm{T}}. \tag{26}$$

The resultant Jacobian matrix of the period-m motion is

$$DP = DP_{mN(mN-1)\ldots1} = \left[\frac{\partial \mathbf{y}_{mN}^{(m)}}{\partial \mathbf{y}_0^{(m)}}\right]_{(\mathbf{y}_0^{(m)*}, \mathbf{y}_1^{(m)*}, \ldots, \mathbf{y}_{mN}^{(m)*})}$$

$$= \mathbf{A}_{mN}^{(m)} \mathbf{A}_{mN-1}^{(m)} \dots \mathbf{A}_1^{(m)} = \mathbf{A}^{(m)}, \tag{27}$$

where

$$\Delta \mathbf{y}_{mN}^{(m)} = \mathbf{A}^{(m)} \Delta \mathbf{y}_0^{(m)} = \mathbf{A}_{mN}^{(m)} \mathbf{A}_{mN-1}^{(m)} \dots \mathbf{A}_1^{(m)} \Delta \mathbf{y}_0^{(m)},$$

$$\Delta \mathbf{y}_k^{(m)} = \mathbf{A}_k^{(m)} \Delta \mathbf{y}_{k-1}^{(m)}, \quad \mathbf{A}_k^{(m)} \equiv \left[\frac{\partial \mathbf{y}_k^{(m)}}{\partial \mathbf{y}_{k-1}^{(m)}} \right]_{(\mathbf{y}_{k-1}^{(m)*}, \mathbf{y}_k^{(m)*})}, \tag{28}$$

and

$$\mathbf{A}_k^{(m)} = \begin{bmatrix} \mathbf{B}_k^{(m)} & a_{k(r_{k-1}-1)}^{(m)} \\ \mathbf{I}_k^{(m)} & \mathbf{0}_k^{(m)} \end{bmatrix}_{(s+1) \times (s+1)}, \ s = 1 + l_{k-1},$$

$$\mathbf{B}_k^{(m)} = [a_{k(k-1)}^{(m)}, \underbrace{0, \dots, 0}_{s-3}, a_{kr_k}^{(m)}, a_{kr_{k-1}}^{(m)}],$$

$$\mathbf{I}_k^{(m)} = diag(1, 1, \dots, 1, 1)_{s \times s},$$

$$\mathbf{0}_k^{(m)} = \underbrace{(0, 0, \dots, 0, 0)}_{s}^{\mathrm{T}};$$

$$a_{kj}^{(m)} = -(\frac{\partial g_k}{\partial x_k^{(m)}})^{-1} \frac{\partial g_k}{\partial x_j^{(m)}},$$

$$a_{kr_j}^{(m)} = -(\frac{\partial g_k}{\partial x_k^{(m)}})^{-1} \sum_{\alpha=j}^{j+1} \frac{\partial g_k}{\partial x_\alpha^{\tau(m)}} \frac{\partial x_\alpha^{\tau(m)}}{\partial x_{r_j}^{(m)}},$$

$$a_{k(r_j-1)} = -(\frac{\partial g_k}{\partial x_k^{(m)}})^{-1} \sum_{\alpha=j-1}^{j} \frac{\partial g_k}{\partial x_\alpha^{\tau(m)}} \frac{\partial x_\alpha^{\tau(m)}}{\partial x_{r_j-1}^{(m)}},$$

$$\text{with } r_j = j - l_j, j = k-1, k; \tag{29}$$

and

$$\frac{\partial g_k}{\partial x_k^{(m)}} = 1 - \tfrac{1}{2} h \Delta, \qquad \frac{\partial g_k}{\partial x_{k-1}^{(m)}} = -1 - \tfrac{1}{2} h \Delta,$$

$$\frac{\partial g_k}{\partial x_j^{\tau(m)}} = \tfrac{1}{2} \alpha_2 h \cos M, \quad \frac{\partial x_j^{\tau(m)}}{\partial x_{r_j}^{(m)}} = 1 - \tfrac{1}{h} \tau + l_{r_j},$$

$$\frac{\partial x_j^{\tau(m)}}{\partial x_{r_j-1}^{(m)}} = \tfrac{1}{h} \tau - l_{r_j},$$

$$\Delta = \alpha_1 - \tfrac{3}{4} \beta (x_k^{(m)} + x_{k-1}^{(m)})^2,$$

$$M = \tfrac{1}{2} [x_{k-l_k-1}^{(m)} + x_{k-l_k-2}^{(m)} + (1 - \tfrac{1}{h} \tau + l_k)(x_{k-l_k}^{(m)} - x_{k-l_k-2}^{(m)})]. \tag{30}$$

The eigenvalues of DP for the period-m motion in the 1-D, time-delay, nonlinear dynamical system are computed by

$$|DP - \lambda \mathbf{I}_{(s+1) \times (s+1)}| = 0. \tag{31}$$

(i)   If the magnitudes of all eigenvalues of DP are within the unit circle (i.e., $|\lambda_i| < 1$, $i = 1, 2, \ldots, (s + 1)$), the approximate periodic solution is stable.

(ii) If at least the magnitude of one eigenvalue of DP is falling outside of the unit circle (i.e., $|\lambda_i| > 1$, $i \in \{1, 2, \ldots, (s + 1)\}$), the approximate periodic solution is unstable.

(iii) The boundaries between stable and unstable periodic flow with higher order singularity give bifurcation and stability conditions with higher order singularity.

The bifurcation conditions are given as follows.

(iv) If $\lambda_i = 1$ with $|\lambda_j| < 1$ ($i, j \in \{1, 2, \ldots, (s + 1)\}$ and $i \neq j$), the saddle-node bifurcation (SN) occurs.

(v)   If $\lambda_i = -1$ with $|\lambda_j| < 1$ ($i, j \in \{1, 2, \ldots, (s + 1)\}$ and $i \neq j$), the period-doubling bifurcation (PD) occurs.

(vi) If $|\lambda_{i,j}| = 1$ with $|\lambda_l| < 1$ ($i, j, l \in \{1, 2, \ldots, (s + 1)\}$ and $\lambda_i = \overline{\lambda}_j$ $l \neq i, j$), Neimark bifurcation (NB) occurs.

5 Sequential Periodic Motions

In this section, periodic motions varying with excitation frequency in the 1-dimensional, time-delay, dynamical system will be presented from the periodic node displacement $x_{mod(k,N)}$ for $\mod(k, N) = 0$. The periodic nodes are predicted from implicit mappings for periodic motions, and the stability and bifurcation of periodic motions are analyzed through the eigenvalue analysis. Consider a set of parameters as

$$\alpha_1 = 10.0, \quad \alpha_2 = 5.0, \quad \beta = 10.0, \quad Q_0 = 3.0, \quad \tau = T/4, \tag{32}$$

where $T = 2\pi/\Omega$.

The global view of periodic node displacement $x_{mod(k,N)}$ varying with for excitation frequency $\Omega \in (0, 12.0)$ is presented in Fig. 1a. The zoomed picture of the global view for $\Omega \in (5.52, 7.20)$ is presented in Fig. 1b. The solid and dash curves represent stable and unstable periodic motions, respectively. The bold underline letter '**S**' is for symmetric periodic motions; '**A**' is for asymmetric periodic motions. The acronyms 'P-1' is for period-1 motions, 'P-2' is for period-2 motions, and so on. 'SN' and 'PD' is for saddle-node bifurcation and period-doubling bifurcation, respectively. For saddle-node bifurcations, jumping phenomena or symmetry breaks of period-m motions are observed. For the period-doubling bifurcations, period-2 m motions appears from period-m motions.

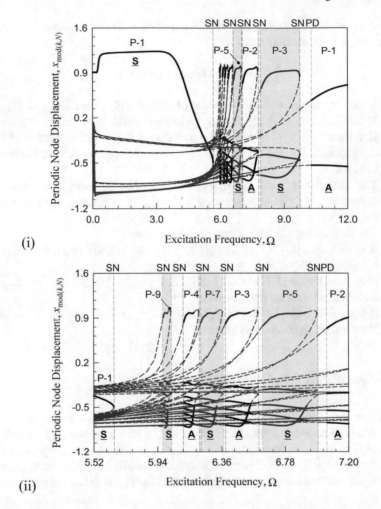

Fig. 1 Node displacement $x_{mod(k,N)}$ varying with excitation frequency for the sequential order of periodic motions: (i) global view for $\Omega \in (0, 12.0)$, (ii) zoomed view for $\Omega \in (5.52, 7.20)$. ($\alpha_1 = 10.0$, $\alpha_2 = 5.0$, $\beta = 10.0$, $Q_0 = 3.0$, $\tau = T/4$). $\mathrm{mod}(k, N) = 0$

In such a range of excitation frequency, there is a global sequence of periodic motions as

$$P_{1(S)} \lhd P_{1(A)} \lhd P_{3(S)} \lhd P_{2(A)} \lhd \ldots \lhd P_{m(A)} \lhd P_{2m+1(S)} \lhd \ldots$$
$$(m = 0, 1, 2, \ldots) \tag{33}$$

where "S" and "A" are for symmetric and asymmetric periodic motions, respectively. Except for the symmetric P-1 motions, all periodic motions in the global sequence have the bifurcation trees to chaos. Herein, focused is the sequences of periodic

Table 1 The bifurcations of asymmetric periodic motions ($\alpha_1 = 10.0$, $\alpha_2 = 5.0$, $\beta = 10.0$, $Q_0 = 3.0$)

Period-m	Frequency range	Ω	Bifurcations	Motion switching
1	$(0, \infty)$	10.285	PD	P1-P2
2	$(0, 7.7577)$	7.7577	SN	J(P2)
		7.748	PD	P2-P4
		7.7048	PD	P2-P4
		7.0573	PD	P2-P4
3	$(0, 6.6003)$	6.6003	SN	J(P3)
		6.599	PD	P3-P6
		6.549	PD	P3-P6
		6.386	PD	P3-P6
4	$(0, 6.2124)$	6.2125	SN	J(P4)
		6.212	PD	P4-P8
		6.18	PD	P4-P8
		6.114	PD	P4-P8

J-for jumping phenomena, SB-for symmetry breaking

motions. The bifurcations of periodic motions in the sequential order are tabulated in Tables 1 and 2.

6 Numerical Illustrations

In this section, periodic motions in the 1-dimensional, time-delay, nonlinear dynamical system will be illustrated through asymmetric period-1 and period-2 motions plus a symmetric period-3 motion. The numerical simulations of periodic motions are carried out by the midpoint scheme, and initial conditions are from analytical predictions. The harmonic amplitudes and phases of periodic motions are provided for effects on periodic motions. Numerical and analytical results are depicted by solid curves and circular symbols, respectively. The initial delay is represented by green circles. The 'D.I.S' is for delay-initial-starting point; and the 'D.I.F' is for delay-initial-finishing point.

Two paired, stable asymmetric period-1 motions are presented at $\Omega = 10.5$ in Fig. 2. The initial conditions are $(x_0, \dot{x}_0) \approx (0.145403, -1.131596)$(black) and $(x_0, \dot{x}_0) \approx (-0.938504, 1.335604)$ (blue). From the finite Fourier series analysis, the harmonic amplitudes and phases are presented in Fig. 2(b) and (c), respectively. The constant is $A_0 = a_0^{(bla)} = -a_0^{(blu)} \approx 0.5356$. The main harmonic amplitudes for the two asymmetric period-1 motions are $A_1 \approx 0.4070$, $A_2 \approx 0.0638$, $A_3 \approx 0.0125$ and $A_4 \approx 3.2285\text{e-}3$.The harmonic amplitude A_1 mainly contributes on the asymmetric period-1 motion, and the asymmetry of the two period-1 motions is contributed from $A_{2l} \neq 0$ terms ($l = 1, 2, \ldots$). Other harmonic amplitudes are

Table 2 The bifurcations of symmetric periodic motions ($\alpha_1 = 10.0, \alpha_2 = 5.0, \beta = 10.0, Q_0 = 3.0$)

Period-m	Frequency range	Ω	Bifurcations	Motion switching
1	(0, 5.6538)	5.6538	SN	J
3	(0, 9.7527)	9.7527	SN	J
		7.912	SN	SB
		7.7764	SN	J
		7.827574	SN	J
		7.827546	PD	P3-P6
5	(0, 6.992911)	6.992911	SN	J
		6.99245	SN	SB
		6.992	PD	P5-P10
		6.9403	PD	P5-P10
		6.9293	SN	SB
		6.6333	SN	SB
		6.6129	PD	P5-P10
		6.114	PD	P4-P8
7	(0, 6.027498)	6.365874	SN	J
		6.36584	SN	SB
		6.36581	PD	P7-P14
		6.331213	PD	P7-P14
		6.331212	SN	J
		6.334354	SN	J
		6.3341	PD	P7-P14
		6.3288	PD	P7-P14
		6.3257	SN	SB
		6.2252	SN	SB
		6.2165	PD	P7-P14
		6.1933	PD	P7-P14
		6.192978	SN	J
		6.2223499	SN	J
		6.22235	PD	P7-P14
9	(0, 6.1054676)	6.1054676	SN	J
		6.105461	SN	SB
		6.1054543	PD	P9-P18
		6.08317	PD	P9-P18
		6.08316	SN	J
		6.0868624	SN	J
		6.08684	PD	P9-P18

(continued)

Table 2 (continued)

Period-m	Frequency range	Ω	Bifurcations	Motion switching
		6.0831695	PD	P9-P18
		6.0801	SN	SB
		6.0332	SN	SB
		6.029	PD	P9-P18
		6.013866	PD	P9-P18
		6.0138554	SN	J
		6.03304044	SN	J
		6.03304042	PD	P9-P18

J-for jumping phenomena, SB-for symmetry breaking

$A_k \in (10^{-15}, 10^{-3})$ $(k = 5, \ldots, 26)$ with $A_{26} \approx 1.6609e\text{-}15$. The harmonic amplitudes drop exponentially with harmonic order. One can use 26 harmonic terms for the accurate approximation of the asymmetric period-1 motions. The harmonic phases for the two paired period-1 motions satisfy $\varphi_k^{(blu)} = \mathrm{mod}\,(\varphi_k^{(bla)} + \pi, 2\pi)$.

Two paired, stable asymmetric period-2 motions are presented at $\Omega = 7.5$ in Fig. 3. The initial conditions are $(x_0, \dot{x}_0) \approx (-0.841663, 1.639460)$(black) and $(x_0, \dot{x}_0) \approx (-0.948285, -2.467868)$ (blue). The two asymmetric period-2 motions still hold $x^{(bla)}(t) = -x^{(blu)}(t + T/2)$ and the two trajectories are of the skew symmetry about the origin. The trajectories of the period-2 motions have one knot. The harmonic amplitudes and phase of the period-2 motions are presented in Fig. 3 (iii) and (iv), respectively. The constants are $A_0 = a_0^{(2)blu} = -a_0^{(2)bla} \approx 0.2563$. The main harmonic amplitudes are $A_{1/2} \approx 0.8104$, $A_1 \approx 0.3791$, $A_{3/2} \approx 0.0446$, $A_2 \approx 0.0922$, $A_{5/2} \approx 0.0295$, $A_3 \approx 0.0158$, $A_{7/2} \approx 0.0128$, $A_4 \approx 7.6258e\text{-}3$, $A_{9/2} \approx 4.5782e\text{-}3$, $A_5 \approx 3.3579e\text{-}3$, $A_{11/2} \approx 2.3988e\text{-}3, A_6 \approx 1.2816e\text{-}3$. The harmonic amplitudes of $A_{1/2} \approx 0.8104$ and $A_1 \approx 0.3791$ contribute on the two period-2 motions. Other harmonic amplitudes are $A_{k/2} \in (10^{-15}, 10^{-3})(k = 13, 14, \ldots, 80)$ and $A_{40} \approx 1.4189e\text{-}15$. With increasing harmonic orders, the harmonic amplitudes decrease. For such a pair of asymmetric period-2 motions, one can use 80 harmonic terms for accurate approximation. The two harmonic phases satisfy $\varphi_{k/2}^{(blu)} = \mathrm{mod}$ $(\varphi_{k/2}^{(bla)} + (k/2 + 1)\pi, 2\pi)$ $(k = 1, 2, \ldots)$.

A stable symmetric period-3 motion is presented at $\Omega = 9.5$ in Fig. 4. The initial condition is $(x_0, \dot{x}_0) \approx (-0.865766, 1.858244)$. The main harmonic amplitudes are $A_{1/3} \approx 0.8709$, $A_1 \approx 0.1852$, $A_{5/3} \approx 0.0739$, $A_{7/3} \approx 0.0103$, $A_3 \approx 4.7013e\text{-}3$, $A_{11/3} \approx 1.7055e\text{-}3$. Other harmonic amplitudes are in $A_{k/3} \in (10^{-15}, 10^{-3})$ $(k = 13, 15, \ldots, 67)$ and $A_{67/3} \approx 1.8603e\text{-}15$. Because $A_{2k/3} = 0$ $(k = 0, 1, \ldots)$, the trajectory of the period-3 motion is symmetric. Over all the harmonic amplitudes decease with harmonic order. One can use 67 harmonic terms to approximate the symmetric period-3 motions.

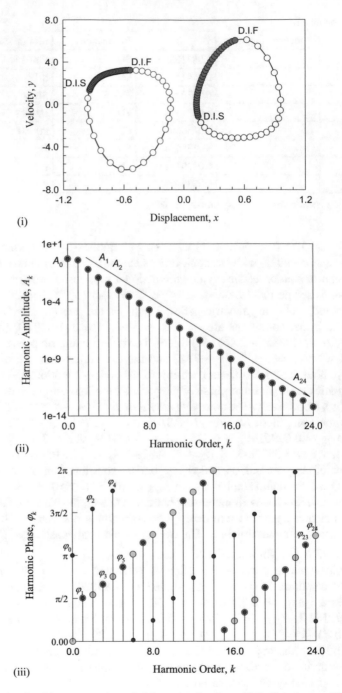

Fig. 2 A pair of stable asymmetric period-1 motion for $\Omega = 10.5$. (i) trajectory (ii) harmonic amplitudes, (iii) harmonic phases. (black $(x_0, \dot{x}_0) \approx (0.145403, -1.131596)$, blue:$(x_0, \dot{x}_0) \approx (-0.938504, 1.335604)$), ($\alpha_1 = 10.0$, $\alpha_2 = 5.0$, $\beta = 10.0$, $Q_0 = 3.0$, $\tau = T/4$)

Fig. 3 Two paired stable asymmetric period-2 motions for $\Omega = 7.5$. (i) trajectory (black): $(x_0, \dot{x}_0) \approx (-0.841663, 1.639460)$, (ii) trajectory (blue):$(x_0, \dot{x}_0) \approx (-0.948285, -2.467868)$, (iii) amplitudes, (iv) phases. $(\alpha_1 = 10.0,\ \alpha_2 = 5.0,\ \beta = 10,\ Q_0 = 3.0,\ \tau = T/4)$ Continued

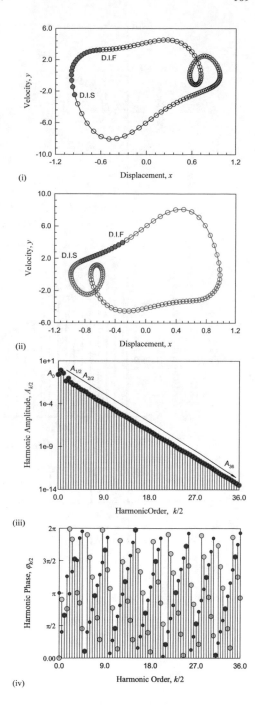

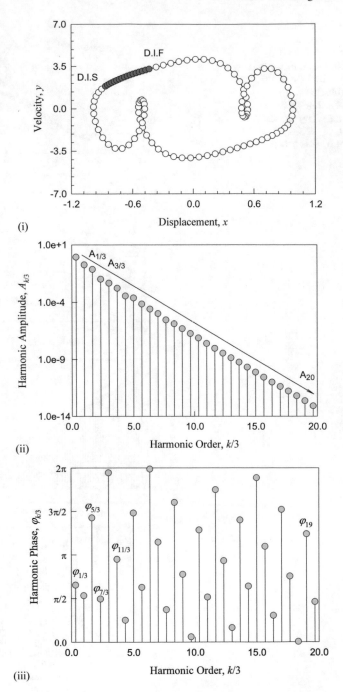

Fig. 4 A stable symmetric period-3 motion for $\Omega = 9.5$. (i) trajectory with, (ii) harmonic amplitudes, (iii) harmonic phases. $(x_0, \dot{x}_0) \approx (-0.865766, \ 1.858244)$ $(\alpha_1 = 10.0, \ \alpha_2 = 5.0, \ \beta = 10.0, \ Q_0 = 3.0, \ \tau = T/4)$

7 Conclusions

The period-m motions in the 1-dimensional, time-delay, nonlinear dynamical system were discussed through the semi-analytical method. A global sequence of periodic motions was discovered. From the analytical prediction, numerical simulations of periodic motions were completed. Harmonic amplitudes and phases for periodic motions were determined by the finite Fourier series analysis. Through such studies, possible complex periodic motions in the 1-dimensional, time-delay, nonlinear dynamical systems were presented.

References

1. Mackey MC, Glass D (1977) Oscillation and chaos in physiological control systems. Science 197(4300):287–289
2. Namajūnas A, Pyragas K, Tamaševičius A (1995) An electronic analog of the Mackey-Glass system. Physics Letter A 201(1):42–46
3. Ikeda K, Daido H (1980) Optical turbulence: chaotic behavior of transmitted light from a ring cavity. Phys Rev Lett 45(9):709–712
4. Ikeda K, Matsumoto K (1987) High-dimensional chaotic behavior in systems with time-delayed feedback. Physica D 29(1–2):223–235
5. Yang JH, Sanjuán MAF, Wang CJ, Zhu H (2013) Vibrational resonance in a Duffing system with a generalized delayed feedback. J Appl Nonlinear Dyn 2(4):397–408
6. Jeevarathinam C, Rajasekar S, Sanjuán MAF (2015) Vibrational resonance in the Duffing oscillator with distributed time-delayed feedback. J Appl Nonlinear Dynam, 4(4), 391–404
7. Krishnaveni V, Sathiyanathan K (2014) Synchronization of couple map lattice using delayed variable feedback. J Appl Nonlinear Dyn 3(3):227–244
8. Akhmet M (2014) Synchronization of the cardiac pacemaker model with delayed pulse-coupling. Discont, Nonlinear Compl 3(1):19–31
9. Hu HY, Dowell EH, Virgin LN (1998) Resonance of harmonically forced Duffing oscillator with time-delay state feedback. Nonlinear Dyn 15(4):311–327
10. Hu HY, Wang ZH (2002) Dynamics of controlled mechanical systems with delayed feedback. Springer, Berlin
11. MacDonald N (1995) Harmonic balance in delay-differential equations. J Sounds Vibr 186(4):649–656
12. Leung AYT, Guo Z (2014) Bifurcation of the periodic motions in nonlinear delayed oscillators. J Vib Control 20:501–517
13. Luo ACJ (2012) Continuous dynamical systems. HEP/L&H Scientific, Beijing/Glen Carbon
14. Luo ACJ, Huang JZ (2012) Analytical dynamics of period-m flows and chaos in nonlinear systems. Int J Bifurcat Chaos 22(1250093): 29
15. Luo ACJ (2013) Analytical solutions of periodic motions in dynamical system with/without time-delay. Int J Dyn Contr 1:330–359
16. Luo ACJ, Jin HX (2014) Bifurcation trees of period-m motion to chaos in a Time-delayed, quadratic nonlinear oscillator under a periodic excitation. Discont Nonlinear Complex 3:87–107
17. Luo ACJ (2015) Discretization and implicit mapping dynamics. HEP/Springer, Beijing/Dordrecht
18. Luo ACJ, Guo Y (2015) A semi-analytical prediction of periodic motions in Duffing oscillator through mapping structures. Discont Nonlinear Complex 4(2):121–150

19. Luo ACJ, Xing SY (2016) Multiple bifurcation trees of period-1 motions to chaos in a periodically forced, time-delayed, hardening duffing oscillator. Nonlinear Dyn 89:405–434
20. Xing SY, Luo ACJ (2017) Towards infinite bifurcation trees of period-1 motions to chaos in a time-delayed, twin-well duffing oscillator. J Vibrat Test Sys Dyn 1(4):353–392
21. Luo ACJ, Xing SY (2017) Time-delay effects on periodic motions in a Duffing oscillator, *Chaotic, Fractional, and Complex Dynamics: New Insights and Perspectives*. Understanding Complex Systems, Springer, Cham
22. Xu Y, Luo ACJ (2018) A series of symmetric period-1 motions to chaos in a two-degree-of-freedom van der Pol-Duffing oscillator. J Vibrat Test Sys Dyn 2(2):119–153

On the Geometric Approach to Transformations of the Coordinates of Inertial Frames of Reference

A. A. Talyshev

Abstract In this paper we consider the problem of constructing coordinate transformations inertial reference systems based solely on geometric considerations. It is shown that within the framework of Euclidean geometry there are only three classes transformations, two of which correspond to the transformations of Galileo and Lorentz.

1 Introduction

The Lorentz transformations were originally obtained by Lorentz in the analysis of the Maxwell equations (see paper [1]). Derivation of the coordinate transformation formulas in the paper [2] and in the subsequent educational literature is based on the following two postulates.

1. The laws of physics are the same for all observers in uniform motion relative to one another (principle of relativity)
2. The speed of light in a vacuum is the same for all observers, regardless of their relative motion or of the motion of the light source.

However, the meaning and application of the theory is not limited to framework of electromagnetism and therefore it is clear the desire to construct a theory based on more general assumptions. In the works [3, 4] it is proposed such a conclusion is based solely on geometric considerations. However, in these works, as well as in [2], it is assumed that it is possible to construct in a moving reference frame line parallel to a line stationary system. This assumption is generally untrue (at least requires specifications), but it is not fully used in the indicated works and therefore allows to achieve success (see Sect. 6).

A. A. Talyshev (✉)
Novosibirsk State University, Pirogova st. 2, Novosibirsk, Russia
e-mail: tal@academ.org

© Higher Education Press 2021
D. Volchenkov (ed.), *Nonlinear Dynamics, Chaos, and Complexity*,
Nonlinear Physical Science, https://doi.org/10.1007/978-981-15-9034-4_8

113

In this paper we consider the problem of constructing all possible linear transformations of coordinates of inertial reference frames of homogeneous isotropic spaces also based on geometric considerations, but with less than [3, 4] number of initial assumptions. It is shown that the required transformations form a one-parameter family, and up to normalization there are three classes of transformations, which, of course, belong to the transformation of Galileo, and Lorentz, and also that of the whole family are most consistent with the available experimental data of the Lorentz transformation.

2   Statement of the problem

We consider the vector space of events $S = T \times P$, where T is one-dimensional, and P is a three-dimensional Euclidean spaces. The metric in the four-dimensional space S is not introduced. The basis in the four-dimensional space S will always be composed of the following vectors: direct product of the basis vector of the space T by the zero vector space P, and the zero vector of the space T by the vectors of some orthonormal basis of the space P. The basic vectors will be denoted by the symbols e_0, e_1, e_2, e_3, and the coordinates of the vectors of the spaces T and P will be denoted by the symbols t and $\mathbf{r} = (x, y, z)$ respectively.

The coordinate representation of the mapping T in P

$$\mathbf{r} = \varphi(t)$$

will be called the equation of motion of a point. For a differentiable of the mapping φ the derivative of the vector-valued function φ with respect to t will be called the speed of motion of the point and be denoted by φ_t.

In this paper, unlike [2], it is proposed to abandon the possibility of direct comparison of there were no objects, in particular bases, in moving systems counting. This leads to a consideration of the relatively each other spaces of events of the kind indicated above (instead of frames of reference). The reasons for this approach will be discussed in the section 6, at least, that the proposed approach does not reduce the generality of the problem under consideration.

Suppose that some basis is chosen in the space S. Under the choice basis in S will be understood not only the choice of the orientation of the vectors in P, but also the choice of metrics in the spaces T and P.

Definition 1 The space $S' = T' \times P'$ moves with respect to the space $S = T \times P$ with the speed $\mathbf{u} \in R^3$, if there exists a one-parameter family of mappings of the space P' in the space P such that the coordinate representation $F(t)$ of this family satisfies the following condition

$$F(t_2)\mathbf{r}' - F(t_1)\mathbf{r}' = \mathbf{u}(t_2 - t_1) \tag{1}$$

for any $\mathbf{r}' \in R^3$ and $t_1, t_2 \in R$. By R we denote the field of real numbers.

Remark 1 So the defined concept of motion is invariant with respect to the choice of bases in spaces S and S', but the velocity of motion depends on the choice of basis in the space S, and the shape of the mapping F depends on the choice of bases in the spaces S and S'.

The mapping specified in the definition partially relates the coordinates of the events the spaces S and S'

$$\mathbf{r} = F(t)\mathbf{r}' = \mathbf{u}t + F(0)\mathbf{r}'. \tag{2}$$

Linear mapping of the coordinates of any four-dimensional spaces S and S' can be written in the form

$$
\begin{aligned}
t' &= \alpha t + \beta \mathbf{r}, \\
\mathbf{r}' &= \gamma t + \theta \mathbf{r},
\end{aligned}
\tag{3}
$$

where $\alpha \in R$, $\beta \in \mathscr{L}(R^3, R)$, $\gamma \in \mathscr{L}(R, R^3)$, $\theta \in \mathscr{L}(R^3, R^3)$. Where $\mathscr{L}(R^k, R^m)$ is the space of linear mappings from R^k to R^m.

If the space S' moves relative to the space S with the speed $\mathbf{u}$, then the coefficients of the map (3) depend on the vector $\mathbf{u}$, and also relations (2) must follow from the relations (3). Below we assume a smooth dependence of the transformations (3) on the velocity $\mathbf{u}$.

So, let the space S' move relative to the space S with the speed $\mathbf{u}$ and their coordinates are related by the relations (3). Then the equation of motion

$$\mathbf{r} = \varphi(t) \tag{4}$$

in the space S there corresponds the equation of motion

$$\mathbf{r}' = \varphi'(t') \tag{5}$$

in the space S', where the map φ' is found from the condition

$$\gamma t + \theta \varphi(t) = \varphi'(\alpha t + \beta \varphi(t)). \tag{6}$$

The differentiation (6) with respect to t gives

$$\gamma + \theta \varphi_t = \varphi'_{t'}(\alpha + \beta \varphi_t),$$

whence

$$\varphi'_{t'} = \frac{\gamma + \theta \varphi_t}{\alpha + \beta \varphi_t}. \tag{7}$$

The spaces S and S' are distinct spaces, nevertheless, here the map (3) will be called the transformation coordinates. Also for a point moving in the space S according to the law (4), it will be talked about its movement relative to space S' exclusively in the sense of the formulas (5), (6). It is in this sense that the formula (7) gives the transformation speed of a moving point.

Resting relative to the S' point should move relative to S with speed $\mathbf{u}$, therefore from the formula (7) follows that

$$\gamma(u) + \theta(\mathbf{u})\mathbf{u} = 0. \tag{8}$$

Substitution (2) in the second of the relations (3) under the condition (8) gives

$$\mathbf{r}' = \gamma(\mathbf{u})t + \theta(\mathbf{u})(\mathbf{u}t + F(0)\mathbf{r}') = \theta(\mathbf{u})F(0)\mathbf{r}',$$

from which it follows that the mapping $F(0)$ is linear and invertible, and also that it is necessary to assume its dependence on the speed $\mathbf{u}$.

The exception of t from the relations (3) gives

$$\mathbf{r}' = \gamma\alpha^{-1}t' + (\theta - \alpha^{-1}\gamma\beta)\mathbf{r},$$

from which it follows that the space S moves with respect to space S' with the speed $\alpha^{-1}\gamma$. The same speed value is obtained from the formula (7).

It is obviously that the coefficients of transformations (3) depend on the choice bases in the spaces S and S'. For further constructions it is useful to eliminate the arbitrariness caused by the choice of bases. For this it is sufficient to indicate the method of choosing a basis in the space S' over the basis in the space S.

In what follows, the basis in S' is chosen in such a way that the speed of motion of the space S relative to the space S' was equal to $-\mathbf{u}$. This leads to the condition

$$\alpha^{-1}\gamma = -\mathbf{u}. \tag{9}$$

This choice fixes the ratio of the scales of the spaces P' and T', and also partially the orientation of the basis in the space P'. The remaining arbitrariness will be eliminated by comparing the basis of the space P' with the image of a basis of the space P under the mapping $F^{-1}(0)$.

Taking into account (9), it is convenient, changing the notation ($\beta \to \alpha\beta$, $\theta \to \alpha\theta$), to write down the transformation (3) in the form

$$\begin{aligned}
t' &= \alpha(\mathbf{u})(t + \beta(\mathbf{u})\mathbf{r}), \\
\mathbf{r}' &= \alpha(\mathbf{u})(-\mathbf{u}t + \theta(\mathbf{u})\mathbf{r}).
\end{aligned} \tag{10}$$

3 Isotropy

When the bases in both spaces P and P' rotate, with the aid of the matrix B' (the prime here denotes transposition) the coordinates of the points and the velocity components are transformed in the following way:

$$\mathbf{r} \to B\mathbf{r}, \quad \mathbf{u} \to B\mathbf{u}, \quad \mathbf{r}' \to B\mathbf{r}'. \tag{11}$$

If the rule for choosing a basis in S' with respect to a basis in S is invariant with respect to identical rotations in the spaces P and P', then the new coordinates (11) must also be connected transformations of the form (10), i.e.

$$t' = \alpha(B\mathbf{u})(t + \beta(B\mathbf{u})B\mathbf{r}),$$
$$B\mathbf{r}' = \alpha(B\mathbf{u})(-B\mathbf{u}t + \theta(B\mathbf{u})B\mathbf{r}). \tag{12}$$

Substituting the values of t' and $\mathbf{r}'$ from (10) into (12) gives

$$\alpha(\mathbf{u})(t + \beta(\mathbf{u})r) = \alpha(B\mathbf{u})(t + \beta(B\mathbf{u})B\mathbf{r}),$$
$$\alpha(\mathbf{u})B(-\mathbf{u}t + \theta(\mathbf{u})r) = \alpha(B\mathbf{u})(-B\mathbf{u}t + \theta(B\mathbf{u})B\mathbf{r}),$$

or

$$\alpha(\mathbf{u}) = \alpha(B\mathbf{u}),$$
$$\beta(\mathbf{u}) = \beta(B\mathbf{u})B, \tag{13}$$
$$B\theta(\mathbf{u}) = \theta(B\mathbf{u})B.$$

For the immediate purposes it is sufficient that the rule of choice of basis in the space S' with respect to the basis in the space S was invariant under identical rotations of the bases in spaces P and P' around the velocity vector. The already established agreement on bases in spaces S and S' this condition is satisfied, and the remaining arbitrariness in the choice basis of the space S', consisting in the possibility rotation of the basis of the space P' around the velocity vector, in the choice of orientation with respect to reflection and scale in P' it is also obvious that this can be eliminated by satisfying this condition.

If the bases satisfy the above convention and $B\mathbf{u} = \mathbf{u}$, then the transformation coefficients, as follows from (13), must satisfy the following conditions:

$$\beta(\mathbf{u}) = \beta(\mathbf{u})B,$$
$$B\theta(\mathbf{u}) = \theta(\mathbf{u})B. \tag{14}$$

Lemma 1 *The conditions (14) are satisfied if and only if the actions of the matrices β and θ can be written in the form*

$$\beta(\mathbf{u})\mathbf{r} = \tilde{\beta}(\mathbf{u}, \mathbf{r}),$$
$$\theta(\mathbf{u})\mathbf{r} = \varkappa\mathbf{r} + \sigma(\mathbf{u}, \mathbf{r})\mathbf{u} - \mu(\mathbf{u} \times \mathbf{r}),$$

(15)

where $\tilde{\beta}$, $\varkappa$, σ, μ are some scalar functions of the vector $\mathbf{u}$, and $(\mathbf{u}, \mathbf{r})$ is a scalar and $(\mathbf{u} \times \mathbf{r})$ is a vector products of vectors.

Proof *Necessity.* When both parts are transposed, the matrix equality $\beta = \beta B$ passes into the equality $\beta' = B'\beta'$ or $\beta' = B^{-1}\beta'$, since $B' = B^{-1}$. Thus, $B\beta' = \beta'$, i.e., vector β' is a right eigenvector for any matrix B of the corresponding transformation of rotation around the velocity vector $\mathbf{u}$, and therefore must be proportional to the vector $\mathbf{u}$. So, there is a scalar function $\tilde{\beta}(\mathbf{u})$ such that $\beta' = \tilde{\beta}(\mathbf{u})\mathbf{u}$ and therefore the action of the matrix β can be written in the form $\beta(\mathbf{u})\mathbf{r} = \tilde{\beta}(\mathbf{u})(\mathbf{u}, \mathbf{r})$.

Any vector from R^3 can be decomposed into three noncoplanar vectors. In particular, for every $\mathbf{r}$ of a nonparallel $\mathbf{u}$ there exist scalar functions $\varkappa(\mathbf{u})$, $\eta(\mathbf{u}, \mathbf{r})$, $\mu(\mathbf{u})$ such that

$$\theta(\mathbf{u})\mathbf{r} = \varkappa\mathbf{r} + \eta\mathbf{u} + \mu(\mathbf{u} \times \mathbf{r}).$$

(16)

Here $\varkappa$ and μ do not depend on $\mathbf{r}$, because of the linear dependencies of the left side (16) on $\mathbf{r}$. For the same reason, the function η is linear in $\mathbf{r}$. Since $B\theta\mathbf{u} = \theta B\mathbf{u} = \theta\mathbf{u}$, then $\theta\mathbf{u}$ is parallel to $\mathbf{u}$ and therefore the representation (16) is also valid for the vectors $\mathbf{r}$ parallel to $\mathbf{u}$.

Substitution of the representation (16) into the equality $B\theta\mathbf{r} = \theta B\mathbf{r}$ gives

$$\varkappa(\mathbf{u})B\mathbf{r} + \eta(\mathbf{u}, \mathbf{r})B\mathbf{u} + \mu(\mathbf{u})B(\mathbf{u} \times \mathbf{r}) =$$
$$= \varkappa(\mathbf{u})B\mathbf{r} + \eta(\mathbf{u}, B\mathbf{r})\mathbf{u} + \mu(\mathbf{u})(\mathbf{u} \times B\mathbf{r})$$

and since $B\mathbf{u} = \mathbf{u}$, and $B(\mathbf{u} \times \mathbf{r}) = (\mathbf{u} \times B\mathbf{r})$, then it follows that $\eta(\mathbf{u}, \mathbf{r}) = \eta(\mathbf{u}, B\mathbf{r})$. From this and since η is linear in $\mathbf{r}$ there exists function $\sigma(\mathbf{u})$ such that $\eta(\mathbf{u}, \mathbf{r}) = \sigma(\mathbf{u})(\mathbf{u}, \mathbf{r})$.

The sufficiency is verified by substituting the obtained representations into formulas (14). □

The condition (8) is now written in the form

$$-\mathbf{u} + \varkappa\mathbf{u} + \sigma\mathbf{u}^2\mathbf{u} = 0$$

and since this relation holds for all vectors $\mathbf{u}$ it follows that functions $\varkappa$ and σ satisfy the following condition

$$\varkappa(\mathbf{u}) + \sigma(\mathbf{u})\mathbf{u}^2 = 1.$$

(17)

Now we can refine the rule for choosing a basis in S', for this it is more convenient to consider the transformation for the velocity vector $\mathbf{u} = (u, 0, 0)$

$$\begin{aligned}
t' &= \alpha(t + \tilde{\beta}ux), \\
x' &= \alpha(-ut + x), \\
y' &= \alpha(\varkappa y + \mu uz), \\
z' &= \alpha(-\mu uy + \varkappa z)
\end{aligned} \tag{18}$$

and consider the image of the basis of the space P in P' under the mapping $F^{-1}(0) = \theta$

$$\mathbf{e}_1 \rightarrow \tilde{\mathbf{e}}_1 = \alpha \begin{pmatrix} 1 \\ 0 \\ 0 \end{pmatrix}, \quad \mathbf{e}_2 \rightarrow \tilde{\mathbf{e}}_2 = \alpha \begin{pmatrix} 0 \\ \varkappa \\ -\mu u \end{pmatrix}, \quad \mathbf{e}_3 \rightarrow \tilde{\mathbf{e}}_3 = \alpha \begin{pmatrix} 0 \\ \mu u \\ \varkappa \end{pmatrix}. \tag{19}$$

The length of the vectors $\tilde{\mathbf{e}}_2$ and $\tilde{\mathbf{e}}_3$ depends on the scale in P', therefore, the determination of the lengths of these vectors fixes the choice of the scale in P' (with respect to P). In the future such a choice is assumed of the scale in P' for which the map $F^{-1}(0)$ preserves the length vectors orthogonal to the velocity vector. This leads to the condition

$$\alpha^2(\varkappa^2 + \mu^2 \mathbf{u}^2) = 1. \tag{20}$$

Further, it is assumed that the orientation of the basis in P' coincides with the orientation of the image of the basis of the space P under the mapping $F^{-1}(0)$, therefore

$$\alpha > 0. \tag{21}$$

And the last one is the image of the basis of the space P, according to (19), are rotated relative to the basis of the space P' around the vector $\mathbf{u}$ in the direction from $\mathbf{e}_2'$ to $\mathbf{e}_3'$ by some angle $\varphi(\mathbf{u})u/|u|$, which leads to the condition

$$\varkappa(\mathbf{u}) = \frac{\cos\varphi(\mathbf{u})}{\alpha(\mathbf{u})}, \quad \mu(\mathbf{u}) = \frac{\sin\varphi(\mathbf{u})}{\alpha(\mathbf{u})|u|}.$$

The definition of the angle φ completes the description of the choice of basis space S' in the basis of the space S.

If the basis is chosen so that the angle φ depends only on the length vector $\mathbf{u}$, then the described way of choosing the basis is invariant with respect to any identical rotations in the spaces P and P'. Thus, the coefficients of the coordinate transformation must satisfy the conditions (13).

Lemma 2 *The conditions (13) are satisfied if and only if the conditions (15) are satisfied and the functions $\tilde{\beta}$, $\varkappa$, σ, μ, α depend only on the length of the vector $\mathbf{u}$.*

Proof Substitution of the obtained values for the transformation coefficients in the formula (13) gives that the functions α, $\tilde{\beta}$, $\varkappa$, σ, μ are invariant relative to the commit transformations of rotation over arguments. Whence the required assertion follows. $\qquad\square$

In what follows, the function $\tilde{\beta}$ will be used without a tilde and when writing functions α, β, σ, $\varkappa$, μ if necessary, the argument will be used as an argument not $|\mathbf{u}|$, but simply $\mathbf{u}$.

4 Inverse transformation

It is easy to verify that this method of choosing a basis in S' over a basis in S is reflexive, i.e. a basis of the space S with respect to the basis of the space S' has the same properties. Therefore, the transformation

$$
\begin{aligned}
t &= \alpha(t' - \beta ux'), \\
x &= \alpha(ut' + x'), \\
y &= \alpha(\varkappa y' - \mu uz'), \\
z &= \alpha(\mu uy' + \varkappa z')
\end{aligned}
\tag{22}
$$

must be the inverse of the transformation (18).

Substitution of the values of (t', x') from (18) in the first two relations (22) gives

$$
\begin{aligned}
t &= \alpha^2(t + \beta ux + \beta u^2 t - \beta ux) = \alpha^2(1 + \beta u^2)t, \\
x &= \alpha^2(ut + \beta u^2 x - ut + x) = \alpha^2(1 + \beta u^2)x,
\end{aligned}
$$

i.e.

$$
\alpha(\mathbf{u}) = \frac{1}{\sqrt{1 + \beta u^2}}.
$$

Substitution of the values of (y', z') from (18) in the last two equations (22) leads, taking into account (17) to identities.

5 Transitivity

Now let the spaces S' and S'' move with respect to the space S with the velocities $\mathbf{u}$ and $\mathbf{v}$, respectively. Already received requirements for the transformation give that in this situation the space S'' moves relative to the space S' with constant uniform speed and allow this speed to be found. Therefore, the coordinates of the spaces S' and S'' must also be are connected by a transformation, which, under condition of correctness the choice of bases in S' and S'' should be expressed formula (10), where it is natural to replace $\mathbf{u}$ is the speed of motion of the space S'' relative to the space S'.

According to the formula (7), the space S'' moves relative to space S' with the speed

$$
\mathbf{v}' = \frac{-\mathbf{u} + \varkappa(\mathbf{u})\mathbf{v} + \sigma(\mathbf{u})(\mathbf{u}, \mathbf{v})\mathbf{u} - \mu(\mathbf{u})(\mathbf{u} \times \mathbf{v})}{1 + \beta(\mathbf{u})(\mathbf{u}, \mathbf{v})},
$$

and the space S' moves relative to the space S'' with the speed

$$\mathbf{u}'' = \frac{-\mathbf{v} + \varkappa(\mathbf{v})\mathbf{u} + \sigma(\mathbf{v})(\mathbf{v}, \mathbf{u})\mathbf{v} - \mu(\mathbf{v})(\mathbf{v} \times \mathbf{u})}{1 + \beta(\mathbf{v})(\mathbf{v}, \mathbf{u})}.$$

Obviously, not for any functions β, $\varkappa$, σ, μ bases in the spaces S' and S'' correspond to the above arrangement on the choice of bases. At least, the vector $\mathbf{u}''$ should have been equal to the vector $-\mathbf{v}'$, and this does not occur for any β, $\varkappa$, σ, μ. Therefore, before writing down the transformation for the coordinates of spaces S' and S'', must be changed accordingly, for example, a basis of the space S''. Let this change of basis in S'' leads to the following change of coordinates

$$t'' \to \lambda t'', \quad \mathbf{r}'' \to B\mathbf{r}''. \tag{23}$$

Then we can write

$$\lambda t'' = \alpha(\mathbf{v}')(t' + \beta(\mathbf{v}')(\mathbf{v}', \mathbf{r}')),$$
$$B\mathbf{r}'' = \alpha(\mathbf{v}')(-\mathbf{v}'t' + \mathbf{r}').$$

Substituting here the expressions for the coordinates of the spaces S' and S'' through the coordinates of the space S leads to

$$\begin{aligned}
\lambda\alpha(\mathbf{v})(t + \beta(\mathbf{v})(\mathbf{v}, \mathbf{r})) = \\
= \alpha(\mathbf{v}')\alpha(\mathbf{u})(t + \beta(\mathbf{u})(\mathbf{u}, \mathbf{r}) + \beta(\mathbf{v}')(\mathbf{v}', -\mathbf{u}t + \theta(\mathbf{u})\mathbf{r})), \\
B\alpha(\mathbf{v})(-\mathbf{v}t + \theta(\mathbf{v})\mathbf{r}) = \\
= \alpha(\mathbf{v}')\alpha(\mathbf{u})(-\mathbf{v}'(t + \beta(\mathbf{u})(\mathbf{u}, \mathbf{r})) + \theta(\mathbf{v}')(-\mathbf{u}t + \theta(\mathbf{u})\mathbf{r})).
\end{aligned} \tag{24}$$

Lemma 3 *The conditions (24) are satisfied if and only if the coefficient β is constant.*

Proof The necessity of the assertion of the lemma is manifested already in the particular case, when the vectors $\mathbf{u}$ and $\mathbf{v}$ are parallel. So, let $\mathbf{u}$ and $\mathbf{v}$ be parallel to the axis $\mathbf{e}_1$, i.e. $\mathbf{u} = (u, 0, 0)$, $\mathbf{v} = (v, 0, 0)$, then $\mathbf{v}' = (v', 0, 0)$ and

$$v' = \frac{-u(\varkappa(\mathbf{u}) + \sigma(\mathbf{u})u^2)v}{1 + \beta(\mathbf{u})uv} = \frac{-u + v}{1 + \beta(\mathbf{u})uv}.$$

The first of the conditions (24) is written in this case in the form

$$\lambda\alpha(\mathbf{v}) = \alpha(\mathbf{v}')\alpha(\mathbf{u})(1 - \beta(\mathbf{v}')v'u),$$
$$\lambda\alpha(\mathbf{v})\beta(\mathbf{v})v = \alpha(\mathbf{v}')\alpha(\mathbf{u})(\beta(\mathbf{u})u + \beta(\mathbf{v}')v').$$

The exception from here λ leads to

$$(1 - \beta(\mathbf{v}')v'u)\beta(\mathbf{v})v = \beta(\mathbf{u})u + \beta(\mathbf{v}')v'$$

or
$$\beta(\mathbf{v}')v'(1 + \beta(\mathbf{v})uv) = \beta(\mathbf{v})v - \beta(\mathbf{u})u.$$

The differentiation of the last expression with respect to v for $v = 0$ leads to a differential equation
$$\dot{\beta}(|\mathbf{u}|)|\mathbf{u}| + \beta(|\mathbf{u}|) = \beta(0),$$

integration of which gives

$$\beta(|\mathbf{u}|)|\mathbf{u}| = \beta(0)|\mathbf{u}| + c_1.$$

By the assumption of the continuous dependence of the desired transformation of the velocity in the neighbourhood of the zero velocity, the constant c_1 should equal zero. Thus, $\beta(|\mathbf{u}|) = \beta(0)$, i.e. β is constant.   □

It is easy to verify that for a constant β the lengths of the vectors $\mathbf{v}'$ and $\mathbf{u}''$ are equal and the scales of the bases of the spaces S' and S'' comply with the agreement. Therefore, the correction of the basis in S'' at the requirement transitivity is only in the rotation of the basis by a certain angle, i.e., the matrix B in (23) corresponds to the rotation transformation.

When choosing bases, there remains one free function, it is the angle $\varphi(\mathbf{u})$. The question arises whether it is possible to achieve the choice of this function, that the bases in the spaces S' and S'' immediately correspond to the agreement?

The requirement $\mathbf{u}'' = -\mathbf{v}'$ leads to the condition

$$-\mathbf{v} + \varkappa(\mathbf{v})\mathbf{u} + \sigma(\mathbf{v})(\mathbf{v}, \mathbf{u})\mathbf{v} - \mu(\mathbf{v})(\mathbf{v} \times \mathbf{u}) =$$
$$= -\mathbf{u} + \varkappa(\mathbf{u})\mathbf{v} + \sigma(\mathbf{u})(\mathbf{u}, \mathbf{v})\mathbf{u} - \mu(\mathbf{u})(\mathbf{u} \times \mathbf{v}),$$

or
$$(\varkappa(\mathbf{v}) - 1 + \sigma(\mathbf{u})(\mathbf{u}, \mathbf{v}))\mathbf{u} + (\varkappa(\mathbf{u}) - 1 + \sigma(\mathbf{v})(\mathbf{v}, \mathbf{u}))\mathbf{v}+$$
$$+(\mu(\mathbf{v}) - \mu(\mathbf{u}))(\mathbf{u} \times \mathbf{v}) = 0,$$

in particular, if the vectors $\mathbf{u}$ and $\mathbf{v}$ are perpendicular, then

$$(\varkappa(\mathbf{v}) - 1)\mathbf{u} + (\varkappa(\mathbf{u}) - 1)\mathbf{v} + (\mu(\mathbf{v}) - \mu(\mathbf{u}))(\mathbf{u} \times \mathbf{v}) = 0,$$

It follows that $\varkappa \equiv 1$ and μ is constant. Substitution of the value $\varkappa$ in the formulas (17), (20) gives
$$\sigma = 0, \quad \beta = \mu^2. \tag{25}$$

Thus, the nonnegativity of β is a necessary condition that the bases in the spaces S' and S'' correspond to the arrangement. It is easy to verify that the conditions (25) are also sufficient for this.

6 Conclusion

Thus, we obtain a one-parameter family of transformations, more precisely three classes of different transformations corresponding to the values of the parameter β: $\beta < 0, \beta = 0, \beta > 0$. For $\beta \neq 0$ the specific the value of β depends on the choice of metrics in the spaces P and T and is a world constant, i.e. the same for all systems with bases corresponding to the agreement.

Now, when every possible transformations are constructed (within the framework of Euclidean geometry), we can choose among them the ones using other considerations based on observations.

The case $\beta = 0$ corresponds to the Galilean transformations, $\beta < 0$ corresponds the Lorentz transformations (which take a generally accepted form for $\varphi = 0$). The case $\beta > 0$ has at least two unpleasant properties. The own time of a moving object is with the motionless will go faster, which is inconsistent with observations for muons. Another trouble is that for how much arbitrarily small velocity $\mathbf{v}$ there is always a finite the sequence of event spaces $S_1, \ldots, S_n$, for which for $i = 2, \ldots, n$ the space S_i moves with respect to the space S_{i-1} with velocity not exceeding $\mathbf{v}$, and S_n moves relative to S_1 with infinite speed.

In general, the cases with $\beta = 0$ and $\beta > 0$ can be rejected for many reasons and very important is the invariance Maxwell's equations. Maxwell's equations hold the shape in different inertial frames of reference, if only the law coordinate transformation corresponds to the case $\beta < 0$ and constant c (speed of propagation of electromagnetic waves in vacuum), the one present in the Maxwell equations equal to $\sqrt{-1/\beta}$.

From the results of the previous paragraph, one uncharacteristic property follows transformations with $\beta < 0$, which consists in the fact that the moving the space P' can not be embedded in the fixed P. The classics found a way out of this situation by calling a time variable with the others and already in the four-dimensional continuum with pseudo-Euclidean metric S' is imbedded in S.

Maybe it's useful for something else and besides mathematical convenience, but at least for the derivation of the transformation formulas themselves, as shown here, in this there is no need. It is enough simply to abandon the direct comparison of motionless and moving objects. Moreover, the a priori grounds, except for traditions and everyday considerations, for this comparison no. The fact is that a moving object or even a point for Euclidean space is a phenomenon external and only internal properties not determined, i.e. a moving point is no longer a point of Euclidean space, and something else for him, for example, mapping. Therefore, any measuring instrument used in a given space for exactly the same measurement of a moving object, to go on a 'journey'. However, for a given space it will already be another measuring device.

Now about the remarks made in the introduction to the works [3, 4]. In these works it is assumed explicitly or implicitly that for the motion of the space S' along the axis x of the space S, the axis x' is parallel to the axis x and some simplifications of the form of the matrix transformations based on this assumption. By themselves, these simplifications based on a wrong premise turn out to be correct. However, the situation described in Sect. 5 under this assumption would exclude from consideration the Lorentz transformation. In these more attention is given to coordinates than to bases, and this is fraught with danger. The thing is that by changing the arrangement on bases one can give transformations, generally speaking, any predefined form.

Finally, the definition of the motion of one system relative to the other. In any definition, it can always be distinguished as a component the motion of the 'trace' of the space P' in the space P, and the rest part of the definition will be a concretization of the correspondence of this 'trace' with P'. In this work it is shown that one can do without this concretization.

References

1. Lorentz, H.A. (1904). *Electromagnetic phenomena in a system moving with any velocity smaller than of light*. Proc. Acad. Sci., Amsterdam, 6, 809
2. Einstein A (1905) Zur Eiektrodynamik der beivegter Körper. Ann. Phys. 17:891–921
3. Terlechkii, Ya.P. (1966). *Paradoxes of the theory of relativity.* – M.: Nauka. (in Russian)
4. Mermin ND (1984) Relativity without light. Amer. J. Phys. 52(2):119–124

Corpuscular Models of Information Transfer in a Random Environment

V. V. Uchaikin

Abstract The process of information transfer though a random environment is treated as interaction of information quanta with randomly distributed entities of the medium called atoms. It is considered on the base of the duality principle, involving along with the basic solution of the transfer equation the importance function, obeying the adjoint (in the Lagrange) sense. A few models of random medium are considered, built by means of space-time shift of initially deterministic media, as well as modifying Poisson point field. There are considered Poisson's distributed random inclusions of finite sizes and nonhomogeneous Poisson field with random spatial intensity. The latter model is successfully applied to astrophysics of high-energy cosmic rays, simulating statistical properties of secondary particles, falling on the earth surface as part of an extensive air shower.

1 Introduction

Most methods of diagnostics of various media from living tissues to cosmic environment use propagation of information carriers (heat, sound, light, X-rays and so on) through a medium under investigation. Interacting with its constituents (atoms, molecules, larger formations and heterogeneities) they keep up information on the medium structure and carry it to a measuring device (receiver, detector). Mathematical description of these processes is made on the basis of integro-differential transport equations, which together with methods of their solving form an important section of theoretical physics, called the *physical kinetics*.

The development of nuclear technologies in the middle of the last century motivated the creation and dissemination of precision numerical methods, because even small errors in calculations could lead to a bad consequence. Without exaggeration, it can be said that mankind first encountered such high demands for accuracy, the answer to which has significantly improved both the transport theory and numerical

V. V. Uchaikin (✉)
Ulyanovsk State University, Ulyanovsk, Russia
e-mail: vuchaikin@gmail.com

© Higher Education Press 2021
D. Volchenkov (ed.), *Nonlinear Dynamics, Chaos, and Complexity*,
Nonlinear Physical Science, https://doi.org/10.1007/978-981-15-9034-4_9

analysis as a whole. The inverse problem theory, the perturbation theory, variational methods, Monte Carlo methods, the finite element method, the duality principle and the adjoint function method—this is not a complete list of new concepts that were brought to life by the "nuclear revolution" significantly transformed both the transfer theory and numerical analysis as a whole.

Perhaps, the most essential concept of the modern transport theory is the *duality principle* involving in description of the process along with *basic transport equation* for the *flux of information*, its adjoint counterpart the *adjoint function*, or information *importance*. According to this principle, a linear functional

$$J = (P, \Phi) \equiv \int P(x)\Phi(x)dx \tag{1}$$

of the information field $\Phi(x)$ obeying a linear equation

$$L\Phi(x) = S(x) \tag{2}$$

is numerically equal to the functional (Φ^+, S) of solution $\Phi^+(x)$ of the adjoint (in the Lagrange sense) equation

$$L^+\Phi^+(x) = P(x). \tag{3}$$

The first effective application of this principle to inhomogeneous equations was realized in the *small perturbation theory* [1]–[5], when the perturbation of measured value $\delta J = J - J_0$ caused by replacement $L_0 \to L + \delta L$, was represented in the form,

$$\delta J = -(\Phi_0^+, \delta L\Phi_0), \tag{4}$$

containing only unperturbed basic (Φ_0) and adjoint (Φ_0^+) functions.

Three kinds of problems are reduced to Eq. (4). First is the sensitivity problem: if the operator is decomposed into small pieces, which of them brings the greatest contribution into the result J? Second is the error estimation: if some of the contributions seems to difficult for calculation, what error may be expected in case we ignore it?

The third kind is approximation of solution of a given problem using known information about some close (*unperturbed*) problem. Formula (4) says that the difference between perturbed (say, transport in an inhomogeneous medium) and unperturbed solutions (say, the case of a homogeneous medium), which can be considered as a correction to unperturbed (trial) solution, is created by the equivalent source $S_{eq} = \delta L\Phi_0$. The small-perturbation theory is linear with respect to inhomogeneity. For example, if the latter consists of small localized inhomogeneities (inclusions), we deal with a system of point sources, if the inclusions are distributed in the medium randomly, the problem is reduced to search of the field created by randomly distributed sources (Sect. 8).

The perturbation idea will take a more understandable form when rewriting Eq. (4) as

$$J = J_0 - (\Phi_0^+, \delta L \Phi_0) + ... \tag{5}$$

We recognize here the beginning of a functional series and understand that to increase the accuracy of the result, this series should be continued. It is convenient to do, using the *Green operators* $G \equiv L^{-1}$ formalism [5, 6]. Inverting Eqs. (2)–(3), we get

$$\Phi(x) = GS(x) \tag{6}$$

and

$$\Phi^+(x) = G^+ P(x). \tag{7}$$

Let us mark unperturbed operators by subscript 0, as above. Evidently, the interrelation

$$LG = L_0 G_0 = I \tag{8}$$

takes place (I stands for the *identity operator*). Representing the perturbed operator L as a sum of unperturbed operator L_0 and perturbing one ΔL_0 (which is not necessarily to be small now), acting by G_0 on both sides of equality obtained and taking into account (8) yield

$$G = G_0 - G_0 \Delta L G. \tag{9}$$

Let us rewrite this equation as

$$G = [I + G_0 \Delta L]^{-1} G_0$$

and expanding the first multiplier into power series, we arrive at the infinite series of the perturbation theory:

$$G = \sum_{n=0}^{\infty} (-G_0 \Delta L)^n G_0. \tag{10}$$

Thus, the perturbed functional is of the form,

$$J = (P, GS) = J_0 + \sum_{n=1}^{\infty} (P, [-G_0 \Delta L]^n G_0 S) = J_0 + J_1 + J_2 + ... \tag{11}$$

where

$$J_0 = (P, G_0 S) = (P, \Phi_0),$$

$$J_1 = -(\Phi_0^+, \Delta L \Phi_0),$$

$$J_2 = (\Phi_0^+, \Delta L G_0 \Delta L \Phi_0),$$

and so on. A partial sum

$$J^N \equiv \sum_{n=0}^{N} P\,[-G_0 \Delta L]^n\,G_0 S. \tag{12}$$

of this series is called the *Nth approximation of the perturbation theory*, and the difference $J^{N+1} - J^N$ the *amendment* of Nth order.

2 Concept of Random Media

The notion of a randomly inhomogeneous medium is usually associated with phenomena like turbulence. Emerging in many currents fluids and gases, it manifests itself in irregular changes hydrodynamic and thermodynamic characteristics in space and time. Real currents in nature (air in the Earth's atmosphere, water in rivers and seas, gas in the atmosphere of stars and interstellar nebulae) and technical devices (boilers, pipes, canals, reactors) are turbulent in most cases. Due to, extraordinary irregularity of turbulent flows stochastic methods are used: turbulent dynamic fields are treated as random functions of space and time (random fields).

Turbulence of the atmosphere plays an important role in the processes propagation of electromagnetic (light and radio) waves in it. The mathematical tools developed in [7]–[9]. However, the behavior of cosmic radiation in the atmosphere turbulence was a weak effect, the same level of cosmic rays variations is due to more slow changes in the properties of the atmosphere and near-Earth space.

In nuclear reactors, a noticeable effect is given by the boiling of the liquid and vibration of the solid-state part of the structure—processes are also easy associated with a randomly inhomogeneous medium [10].

Somewhat less obvious is the application of the random model medium to the calculations of radiation transfer in a stationary (not varying with time) inhomogeneous medium. The most important question, to our view, lies in well-founded determination of statistical characteristics of a random field from the initial data. In the case of a turbulent medium, the answer is given by the statistical hydrodynamics, the case of vibration can be investigated by dynamics methods, but what is the source of information about random field in the case when an original medium is time-independent and deterministic medium?

To answer this question, we consider three following examples.

The first example

A point source, placed at the origin into a time-dependent inhomogeneous non-scattering medium, emits along the x-axis Poissonian flow of particles, moving with a constant velocity v and registered in plane $x = d$. The detector is switched on at time $t = 0$ and turns off at the moment $t = T$. Introducing the notation

$$1(t; T) = \begin{cases} 1, & t \in [0, T] \\ 0, & t \bar{\in} [0, T], \end{cases}$$

we write the expression for contribution into the detector reading of the particle, emitted at t, as

$$q_t = \begin{cases} 1(t + d/v; T) & \text{with probability } e^{-\tau(t)}, \\ 0 & \text{with probability } 1 - e^{-\tau(t)}, \end{cases}$$

where

$$\tau(t) = \int_{t}^{t+d/v} \Sigma(v(t' - t), t')v dt'$$

is the optical length of path between emission registration points. Recall, that $\Sigma(x, t)dx$ is the probability for the particle being at point x at time t to be absorbed in element $(x, x + dx)$ of its path. The random variable q_t introduced above is a *stochastic importance* of this particle.

Let $t_1, t_2, \ldots, t_v$ be random moments of particle production, then the random reading of the detector is represented as

$$q = \sum_{i=1}^{v} q_{t_i},$$

with expectation

$$\mathsf{E}q = S \int_{-d/v}^{T-d/v} \bar{q}(t)dt, \tag{13}$$

where

$$\bar{q}(t) = e^{-\tau(t)} \tag{14}$$

is the mean value of the particle importance in time-dependent medium, and

$$S = \bar{v}/T$$

is the mean number of particles emitted per unit time. Let us transform Eqs. (13)–(14) as follows:

$$\mathsf{E}q = \frac{\bar{v}}{T} \int_{-d/v}^{T-d/v} e^{-\tau(t)}dt = \int_{0}^{\infty} \bar{v}e^{-\tau'}P_\tau(\tau')d\tau'. \tag{15}$$

The function

$$P_\tau(\tau') = \frac{1}{T} \int\limits_{-d/v}^{T-d/v} \delta[\tau(t) - \tau']dt. \tag{16}$$

can be interpreted as a probability density for random variable τ, and (15) as a result of averaging over this variable, which will be market with angular brackets:

$$Eq = \langle J(\tau) \rangle, \tag{17}$$

$$J(\tau) = \bar{v} e^{-\tau} \tag{18}$$

Expression (18) describes the average reading of the detector under condition, that the optical thickness of the layer is equal to τ, but in (17) τ is interpreted as a random variable. Since τ reflects the properties of random environment, the set of its values can be interpreted as a statistical ensemble with probabilistic measure defined by the density (16). Now there is no time variable, it is hidden in density (16), and formally we deal with an ensemble of time-independent media, whose properties, however, are determined by time-dependence of $\tau(t)$. It is clear from (16) that

$$P\{\tau' \in \Delta\tau\} = \int\limits_{\tau}^{\tau+\Delta\tau} P_\tau(\tau')d\tau' = \frac{1}{T} \int\limits_{-d/v}^{T-d/v} \left[\int\limits_{\tau}^{\tau+\Delta\tau} \delta(\tau(t) - \tau')d\tau' \right] dt = \frac{\Delta T}{T},$$

i.e. probability that the optical thickness of a random realization of the medium occurs in the interval $\Delta\tau$ is equal to the fraction of time $\Delta T/T$, during which is the optical thickness of a non-stationary deterministic medium belongs to the specified interval.

The second example

The second example relates to a time-independent medium, inhomogeneous in all three directions, but this time we will consider not point but a surface source uniformly distributed over a square $[0, Y] \times [0, Z]$ in the plane $x = 0$ with the density S. In this case, instead of (15)–(16) we will have

$$Eq = \bar{v} \frac{1}{YZ} \int\limits_0^Y \int\limits_0^Z e^{-\tau(y,z)} dy\, dz = \int\limits_0^\infty J(\tau') P_\tau(\tau')d\tau', \tag{19}$$

where

$$P_\tau(\tau') = \frac{1}{YZ} \int\limits_0^Y \int\limits_0^Z \delta[\tau(y, z) - \tau']dy\, dz. \tag{20}$$

Again, we obtain a statistical ensemble of media, whose properties are uniquely determined by the function of two variable $\tau(y, z)$ in the square $[0, Y] \times [0, Z]$. As in the first case, the ensemble is obtained by means of random shifts of the initial medium, but now these shifts are in space, not in time. Each sample of the ensemble is characterized by the shift vector $\boldsymbol{\rho}$, perpendicular to the axis OX, the end of which is uniformly distributed over above-indicated square.

The third example

The point isotropic source is placed at the origin, and the detector measures the number of particles intersecting a sphere of radius R in a non-uniform, time-independent non-scattering medium. In this case

$$
Eq = \frac{\bar{\nu}}{4\pi} \int e^{-\tau(\boldsymbol{\Omega})} d\boldsymbol{\Omega} = \int\limits_{0}^{\infty} J(\tau') P_\tau(\tau') d\tau',
$$

where

$$
P_\tau(\tau') = \frac{1}{4\pi} \int \delta[\tau(\boldsymbol{\Omega}) - \tau'] d\boldsymbol{\Omega}.
$$

The statistical ensemble of random medium realizations is obtained here by random rotation of the original inhomogeneous medium relatively to the origin. One can imagine the following picture: particles emitted from the source in random directions $\boldsymbol{\Omega}_1, \boldsymbol{\Omega}_2, \ldots, \boldsymbol{\Omega}_\nu$. Each of them has to pass a layer with respect optical thickness $\tau(\boldsymbol{\Omega}_1), \tau(\boldsymbol{\Omega}_2), \ldots \tau(\boldsymbol{\Omega}_\nu)$ and if we choose for each of them the x-axis along the direction of its motion, then we find that all the particles move along one coordinate axis, but each in its own environment. This is the ensemble of realizations of this random media.

All three examples discussed above demonstrate the same idea: transforming a deterministic medium into random one. It is called the *randomization*. Before going more deep into the idea , we draw the reader attention to the fact that in each of the above considered examples

(1) there is a certain parameter $[t, \boldsymbol{\rho}(y, z), \boldsymbol{\Omega}]$, associated with the production of particles in the source;
(2) the detector sensitivity function does not depend on this parameter, which in homogeneous stationary medium determines independence of stochastic importance from it;
(3) the desired result has the form of integral with respect to this parameter, which can be interpreted as averaging over a random variable associated with realizations of random environment.

3   The Shift-Randomization

Consider the procedure of randomization in a more general statement: as the initial problem, we take the transport of particles in the scattering inhomogeneous time-dependent medium, the interaction of particles with which is described by the linear interaction coefficients $\sum(\mathbf{s}, \mathbf{p})$ and $\sum(\mathbf{s}, \mathbf{p} \to \mathbf{p}')$, where $\mathbf{s} \equiv (\mathbf{r}, t)$, $\mathbf{p} \equiv (\mathbf{\Omega}, E)$. The average value of the readings of the additive detector is expressed in terms of the density of the sources $S(\mathbf{s}, \mathbf{p})$ and the sensitivity function of the detector $D(\mathbf{s}, \mathbf{p})$ by the integral

$$J = \int d\mathbf{s} \int d\mathbf{p} \int d\mathbf{s}' \int d\mathbf{p}' D(\mathbf{s}, \mathbf{p}) G(\mathbf{s}, \mathbf{p}; \mathbf{s}', \mathbf{p}') S(\mathbf{s}', \mathbf{p}'), \qquad (21)$$

where $G(\mathbf{s}, \mathbf{p}; \mathbf{s}', \mathbf{p}')$ is the Green's function of the transport equation. Note that

$$\int d\mathbf{s} \int d\mathbf{p} D(\mathbf{s}, \mathbf{p}) G(\mathbf{s}, \mathbf{p}; \mathbf{s}', \mathbf{p}') = \Phi^+(\mathbf{s}', \mathbf{p}')$$

is the linear deterministic importance in non-homogeneous time-dependent medium.
    Assume that the variables in the source density are separable:

$$S(\mathbf{s}, \mathbf{p}) = S_0(\mathbf{s}) S'(\mathbf{p}), \qquad (22)$$

and what is more

$$\int S_0(\mathbf{s}) d\mathbf{s} = 1. \qquad (23)$$

Substituting (22) into (21) yields

$$J = \int d\mathbf{p} \int d\mathbf{p}' J(\mathbf{p}, \mathbf{p}') S'(\mathbf{p}'), \qquad (24)$$

where

$$J(\mathbf{p}, \mathbf{p}') = \int d\mathbf{s} \int d\mathbf{s}' D(\mathbf{s}, \mathbf{p}) G(\mathbf{s}, \mathbf{p}; \mathbf{s}', \mathbf{p}') S_0(\mathbf{s}'). \qquad (25)$$

To explain further transformations of this expression, we give some trivial, but logically necessary considerations regarding space-time dependence Green's function. Let us shift the points $\mathbf{s}$ and $\mathbf{s}'$, indicated in its arguments, on the same vector $\boldsymbol{\sigma}$ by writing the corresponding transformation the function itself in the form

$$G(\mathbf{s}, \mathbf{p}; \mathbf{s}', \mathbf{p}') \to G(\mathbf{s} + \boldsymbol{\sigma}, \mathbf{p}; \mathbf{s}' + \boldsymbol{\sigma}, \mathbf{p}').$$

We now return the points $\mathbf{s}$ and $\mathbf{s}'$ to their starting positions and perform another transformation: by the vector $-\boldsymbol{\sigma}$ (that is, by the same distance, but in the opposite side) we will move the medium itself:

$$G(s, \mathbf{p}; s', \mathbf{p}') \rightarrow G^{(-\sigma)}(s, \mathbf{p}; s', \mathbf{p}')$$

Here $G^{(-\sigma)}$ denotes the Green function in this new medium. When an infinite homogeneous stationary medium the Green's function invariant on each of these transformations, but in the general case this, of course, not this way. However, in the general case of an inhomogeneous time-dependent medium, both these the transformations are equivalent:

$$G(\mathbf{s} + \sigma, \mathbf{p}; \mathbf{s}' + \sigma, \mathbf{p}') = G^{(-\sigma)}(\mathbf{s}, \mathbf{p}; \mathbf{s}', \mathbf{p}')$$

Inserting a special form of this interrelation

$$G(\mathbf{s}, \mathbf{p}; \mathbf{s}', \mathbf{p}') = G^{(-\mathbf{s}')}(\mathbf{s}, \mathbf{p}; 0, \mathbf{p}') \qquad (26)$$

into Eq. (25), after the change of the integration variable $\mathbf{s} - \mathbf{s}' \rightarrow \mathbf{s}$ we get:

$$J(\mathbf{p}, \mathbf{p}') = \int d\mathbf{s} \int d\mathbf{s}' D^{(-\mathbf{s}')}(\mathbf{s}, \mathbf{p}) G^{(-\mathbf{s}')}(\mathbf{s}, \mathbf{p}; 0, \mathbf{p}') S_0(\mathbf{s}'), \qquad (27)$$

where

$$D^{(-\mathbf{s}')}(\mathbf{s}, \mathbf{p}) = D(\mathbf{s} + \mathbf{s}', \mathbf{p}). \qquad (28)$$

Because of non-negativity of $S_0(\mathbf{s})$ and condition (23), the inner integral in (27) can be treated as a result of averaging over random parameter $\mathbf{s}'$.
Returning to Eq. (24), we have

$$J = \int \left\langle d\mathbf{p} \int d\mathbf{s} \tilde{D}(\mathbf{s}, \mathbf{p}) \tilde{G}(\mathbf{s}, \mathbf{p}; 0, \mathbf{p}') \right\rangle S'(\mathbf{p}') d\mathbf{p}', \qquad (29)$$

where $\tilde{D}$ and $\tilde{G}$ are interpreted now as random functions, or more exactly, random fields, are obtained from the origin medium by shifts it on random vector with distribution density $S_0(\mathbf{s}')$, given by the space-time part of the source density. Introducing designation

$$\tilde{\Phi}^+(\mathbf{s}, \mathbf{p}) = \int d\mathbf{p}' \int d\mathbf{s}' \tilde{D}(\mathbf{s}', \mathbf{p}') \tilde{G}(\mathbf{s}', \mathbf{p}'; \mathbf{s}, \mathbf{p}),$$

we rewrite (29) in the form

$$J = \int \langle \tilde{\Phi}^+(0, \mathbf{p}) \rangle S'(\mathbf{p}) d\mathbf{p} = \int \int \langle \Phi^+(0, \mathbf{p}) \rangle S_\delta(\mathbf{s}, \mathbf{p}) d\mathbf{s} d\mathbf{p}, \qquad (30)$$

where

$$S_\delta(\mathbf{s}, \mathbf{p}) = \delta(\mathbf{s}) S(\mathbf{p}')$$

stands for the distribution density for point instantaneous source.

The function Φ^+ can be interpreted as the importance of an information "qwantum" starting from point x of some fixed realization of the medium and averaged over all possible trajectories with the same beginning in the same realization of the medium. values in the random realization of the environment (we will call it the *semi-stochastic importance*), and the detector reading J as the result of averaging over statistical ensemble of media. The physical meaning of such an interpretation is in that the space-time shift transformation move all trajectories with different initial points in such a way, that they have a common initial point, but each of them passes through its own realization of medium.

In other words, the origin of coordinates for each trajectory coincides now with the point of its birth (random), while the space-time dependence of the environment properties (density, interaction coefficients and so on) of the same medium "looks" for each trajectory differently. Thus, the resulting medium becomes random. In a homogeneous stationary medium, the Green's function remains deterministic, the randomness of the function $\tilde{\Phi}^+$ is now due only to the stochasticity of the detector, which disappears if the sensitivity function (28) does not depend on the vector $\mathbf{s}'$ in the domain of its change determined by density $S_0(\mathbf{s}')$. The case of a stationary inhomogeneous medium was considered in [11]. Further generalization of this approach may include turning transformation and other transformations systems of independent variables.

4   Importance in Quasistationary Environment

Now, we will show in more detail, how change in the medium under diagnostics during the measurement time again generates a statistical ensemble characterizing a special kind of a random medium. We restrict ourselves by consideration of slow changes of the medium and consider a time-dependent random environment. We will call it *quasi-stationary*, if its properties change in time slowly as compared with process of particle propagation from source to detector. If the detector is switched on for some time interval T (measurement time), within which its sensitivity function does not depend on t, and outside of T it is equal to zero, then the indication Such a detector in a quasistationary medium can be written in the form

$$q_T = \sum_i q_{\mathbf{x}_i, t_i} 1(t_i, T),$$

where $\mathbf{x}$ is the state of the particle at the moment t;

$$1(t, T) = \begin{cases} 1, \ t \in T, \\ 0, \ t \overline{\in} T \\ , \end{cases}$$

is the indicator function of measuring interval; $q_{\mathbf{x}_i,t_i}$ is the detector reading caused by a particle with coordinates $\mathbf{x}_i$, t_i. In this case, the stochastic importance $q_{\mathbf{x}_i,t_i}$ depends only on $\Sigma(\mathbf{x}, t)$, characterizing the state medium at time t (spatial density distribution, temperature or other characteristics of the medium). Values of value are in this case, by the functionals of $\Sigma(\mathbf{x}, t)$ as a function of $\mathbf{x}$ under given t, serving as a parameter:

$$\bar{q}_T(\mathbf{x}, t) = \Phi_1(\mathbf{x}_1; \Sigma(\cdot, t))1(t, T), \tag{31}$$

$$\overline{q^2}_T(\mathbf{x}, t) = \Phi_2(\mathbf{x}; \Sigma(\cdot, t))1(t, T), \tag{32}$$

Corresponding moments of detector's reading under condition $S_2(\mathbf{x}_1, t_1; \mathbf{x}_2, t_2) = S(\mathbf{x}_1, t_1)S(\mathbf{x}_2, t_2)$ are written as

$$\bar{q}_T = T \left\langle \int S(\mathbf{x}, t)\Phi_1(\mathbf{x}; \Sigma(\cdot, t))dx \right\rangle_T, \tag{33}$$

and

$$\mathsf{D}q_T = T \left\langle \int S(\mathbf{x}, t)\Phi_2(\mathbf{x}; \Sigma(\cdot, t))dx \right\rangle_T, \tag{34}$$

where symbol

$$\langle \ldots \rangle_T \equiv \frac{1}{T} \int_T \ldots dt$$

denotes time-averaging during T. Writing the source density in the form

$$S(\mathbf{x}, t) = \langle S(\mathbf{x}, t) \rangle_T + \Delta S(\mathbf{x}, t),$$

we obtain instead of (33)–(34):

$$\bar{q}_T = T \left\{ \int \langle S(\mathbf{x}, t) \rangle_T \langle \Phi_1(\mathbf{x}; \Sigma(\cdot, t)) \rangle_T dx + \int \langle \Delta S(\mathbf{x}, t)\Delta \Phi_1(\mathbf{x}; \Sigma(\cdot, t)) \rangle_T dx \right\}, \tag{35}$$

$$\mathsf{D}q_T = T \left\{ \int \langle S(\mathbf{x}, t) \rangle_T \langle \Phi_2(\mathbf{x}; \Sigma(\cdot, t)) \rangle_T dx + \int \langle \Delta S(\mathbf{x}, t)\Delta \Phi_2(\mathbf{x}; \Sigma(\cdot, t)) \rangle_T dx \right\}. \tag{36}$$

Introduce designation $s(\mathbf{x}) = \langle \Sigma(\mathbf{x}, t) \rangle_T$ and expand functionals $\langle \Phi_n \rangle$ ($n = 1, 2$), contained in (35)–(36), in variational series with respect to $\Delta \Sigma(\mathbf{x}, t) = \Sigma(\mathbf{x}, t) - s(\mathbf{x})$:

$$\langle \Phi_n(\mathbf{x}; \Sigma(\cdot, t)) \rangle_T = \Phi_n(\mathbf{x}; s(\cdot)) + \sum_{k=1}^{\infty} \frac{1}{n!} \int d\mathbf{x}_1 \ldots \int d\mathbf{x}_k \times$$

$$\left[\delta^k \Phi_n(\mathbf{x}; s(\cdot))/\delta s(\mathbf{x}_1) \ldots \delta s(\mathbf{x}_n) \right] \langle \Delta \Sigma(\mathbf{x}_1, t) \ldots \Delta \Sigma(\mathbf{x}_k, t) \rangle_T. \tag{37}$$

Transformed expression

$$\langle \Delta\Sigma(\mathbf{x}_1, t) \ldots \Delta\Sigma(\mathbf{x}_k, t) \rangle_T = \int d\Sigma_1 \ldots \int d\Sigma_k (\Sigma_1 - s(\mathbf{x}_1)) \times \ldots$$

$$(\Sigma_k - s(\mathbf{x}_k)) \langle \delta(\Sigma_1 - \Sigma(\mathbf{x}_1, t)) \ldots \delta(\Sigma_k - \Sigma(\mathbf{x}_k, t)) \rangle_T = \int d\Sigma_1 \times \ldots$$

$$\int d\Sigma_k \Delta\Sigma_1 \ldots \Delta\Sigma_k p_k(\Sigma_1, \ldots, \Sigma_k; \mathbf{x}_1, \ldots, \mathbf{x}_k) = \langle \Delta\Sigma_1 \ldots \Delta\Sigma_k \rangle, \qquad (38)$$

shows, that the time average can be interpreted as an average over ensemble of the randomly-inhomogeneous medium (which is marked by angle brackets without subscript T) Probabilistic characteristics of the random function $\Sigma(\mathbf{x})$ of this medium are determined by a set of probability densities

$$p_k(\Sigma_1, \ldots, \Sigma_k; \mathbf{x}_1, \ldots, \mathbf{x}_k) = \langle \delta(\Sigma_1 - \Sigma(\mathbf{x}_1, t)) \ldots \delta(\mathbf{x}_k - \Sigma(\mathbf{x}_k, t)) \rangle_T, \qquad (39)$$

which, in turn, are uniquely determined by the deterministic function of time $\Sigma(bfx, t)$. Indeed, integrating (39) over the intervals $\Delta\Sigma_1, \ldots, \Delta\Sigma_k$,

$$\int_{\Delta\Sigma_1} \ldots \int_{\Delta\Sigma_k} p_k(\Sigma_1, \ldots, \Sigma_k; \mathbf{x}_1, \ldots, \mathbf{x}_k) d\Sigma_1 \ldots d\Sigma_k =$$

$$\langle 1(\Sigma(\mathbf{x}_1, t); \Delta\Sigma_1) \ldots 1(\Sigma(\mathbf{x}_k, t); \Delta\Sigma_k) \rangle_T = T(\Sigma(\mathbf{x}_1, t) \in \Delta\Sigma_1, \ldots, \Sigma(\mathbf{x}_k, t) \in \Delta\Sigma_k)/T,$$

we obtain the fraction of time that the medium spends in state $\Sigma(\mathbf{x}_1, t) \in \Delta\Sigma_1, \ldots,$ $\Sigma(\mathbf{x}_k, t) \in \Delta\Sigma_k$ during measuring time, thus $p_k(\Sigma_1, \ldots, \Sigma_k) d\Sigma_1 \ldots d\Sigma_k$ can be treated as a probability of the event

$$\{\Sigma(\mathbf{x}_1) \in d\Sigma_1, \ldots, \Sigma(\mathbf{x}_k) \in d\Sigma_k\}.$$

Inserting (38) into (37) and rolling up the functional series, we obtain

$$\langle \Phi_n(\mathbf{x}; \Sigma(\cdot, t)) \rangle_T = \langle \Phi_n(\mathbf{x}; \Sigma(\cdot, \mathbf{x})) \rangle.$$

Thus, we arrive at a new formulation of the problem under consideration: Instead of a quasistationary one, we have a randomly inhomogeneous medium, instead of averaging over time—averaging over a statistical ensemble stationary (not changing with time) media.

In its formulation, the study of cosmic ray variations in the Earth's atmosphere— a typical example of an inverse problem: from the results of measurements it is necessary to restore characteristics of the spectrum of the primary radiation $S(\mathbf{x}, t)$ and the medium $\Sigma(x, t)$, in which the radiation is transferred. In the traditional approach as, a link between these quantities the functional is form

$$\bar{q}(t) = \int S(\mathbf{x}, t)\bar{q}_0(\mathbf{x}; \Sigma(\cdot, t))d\mathbf{x},$$

expressing the mathematical expectation of the instantaneous detector through density source S at the time of measurement t and the corresponding quasistationary value $\bar{q}_0$.

The mean values of the readings of the real and instantaneous detectors are related by obvious relations

$$\bar{q}_T = \int_T \bar{q}(t)dt = T\langle\bar{q}(t)\rangle_T, \quad \bar{q}(t) = \lim_{T\to\infty}(\bar{q}_T/T).$$

The difference between the experimentally measured value of q_T/T and the theoretical value $\bar{q}(t)$

$$\Delta = q_T/T - \bar{q}(t) \tag{40}$$

can be of fundamental importance in the sense of the possibility itself restore the required functions and, in any case, determine the error restored characteristics of the source and medium.

Representing the random variable (40) in the form

$$\Delta = [q_T/T - \bar{q}_T/T] + [\bar{q}_T/T - \bar{q}(t)],$$

we find its mean value

$$\bar{\Delta} = \bar{q}_T/T - \bar{q}(t) \tag{41}$$

and the mean square

$$\overline{\Delta^2} = \mathsf{Var}q_T + \overline{[\bar{q}_T/T - \bar{q}(t)]^2}. \tag{42}$$

Representing the random variable (96) in the form

$$\Delta = [q_T/T - \bar{q}_T/T] + [\bar{q}_T/T - \bar{q}(t)],$$

its mean

$$\bar{\Delta} = \bar{q}_T/T - \bar{q}(t)$$

and the mean square

$$\langle\bar{\Delta}\rangle_T = \bar{q}_T/T - \langle\bar{q}(t)\rangle_T = 0,$$

$$\langle\overline{\Delta^2}\rangle_T = \mathsf{Var}_{\text{stat}} + \mathsf{Var}_{\text{dyn}},$$

$$\mathsf{Var}_{\text{stat}} = (\overline{q^2}_T - \bar{q}_T^2)/T \tag{43}$$

is a component of the variance due to statistical fluctuations, and

$$\text{Var}_{\text{dyn}} = \langle (\bar{q}(t))^2 \rangle_T - \langle \bar{q}(t) \rangle_T^2 \tag{44}$$

is dynamical component, vanishing for a stationary source and medium. Accordingly, the square of the relative The measurement error has the form of a sum of two terms

$$\delta^2 = \langle \overline{\Delta^2} \rangle_T / \langle \bar{q} \rangle_T^2 = \delta^2_{\text{stat}} + \delta^2_{\text{dyn}}, \tag{45}$$

in different ways depending on the measurement duration T: δ^2 decreases, but δ^2_{dyn} increases (in the nonstationary case) as T grows. It makes sense to raise the question of determining the optimal measurement time T_0, when the total error is minimal.

The time series of the expressions (35)–(36), which enter into the cumulative error (45), it can be shown that when $T \to 0$, $\delta^2_{\text{stat}} \sim T^{-1}$, $\delta^2_{\text{dyn}} \sim T^2$. Writing (45) as

$$\delta^2 = f(T)/T + T^2 2\, g(T),$$

we get an equation for the optimal time T_0:

$$- f(T_0)/T_0^2 + f'(T_0)/T_0 + 2T_0 g(T_0) + g'(T_0)T_0^2 = 0. \tag{46}$$

In the stationary case, when S and Σ do not depend on time, $f(T) = \text{const}$, and $g(T) = 0$, so that $T_0 = \infty$, since the dynamic component of the error vanishes. In the case of slow changes, it follows from (89) that

$$T_0 \sim [f(0)/2g(0)]^{1/3}.$$

The calculation of the constants $f(0)$, $g(0)$ entering here, as well as a more precise definition The optimal time from equation (46) is possible on the basis of numerical methods.

5   Stochastic Importance in Random Environment

Let us leave instructive examples and pass to consideration of the problem in a more general framework. Stochastic importance of the particle at point x of its phase space depends, except of x, both on the medium realization ω and on the realization of trajectory γ in it. Denoting this function by $q_x(\gamma, \omega)$, we express its expected value through the joint measure $P(d\gamma, d\omega)$ via the double integral

$$\mathsf{E} q_x = \iint q_x(\gamma, \omega) P(d\gamma, d\omega). \tag{47}$$

Assuming, that the environment has a significant effect on statistical properties of particle trajectories, whereas the inverse effect (of particles on the medium) is negligible, it is convenient to introduce a conditional probability measure

$$P(d\gamma, d\omega) = P(d\gamma|\omega)P(d\omega), \tag{48}$$

and rewrite (31) in the form

$$\mathsf{E}q_x = \iint \bar{q}(x, \omega)P(d\omega), \tag{49}$$

where

$$\bar{q}(x, \omega) = \int q_x(\gamma, \omega)P(d\gamma|\omega) \equiv \Phi^+(x, \omega) \tag{50}$$

is the partially averaged stochastic importance: being averaged over trajectory realizations in a fixed environment realization, it remains random because of randomness of the environment itself. We will call (50) the *semi-stochastic importance*. Thus,

$$\mathsf{E}_x q_x = \langle \bar{q}(x) \rangle = \langle \Phi^+(x) \rangle. \tag{51}$$

We see here a result of double averaging: the bar q denotes averaging over all possible trajectories in a fixed realization of the environment, whereas brackets mean its average over all possible realizations of the environment (in case of Monte Carlo simulations such procedures being performed in inverse order, are known as the *double randomization* [30]).

Consider now the variance of q_x:

$$\mathsf{Var}q_x = \mathsf{E}q_x^2 - (\mathsf{E}q_x)^2. \tag{52}$$

In accordance with the above, we have

$$\mathsf{E}_x q_x^2 = \int \overline{q^2}(x, \omega)P(d\omega) = \langle \overline{q^2}(x) \rangle = \langle \mathsf{Var}'q_x \rangle + \langle \bar{q}^2(x) \rangle, \tag{53}$$

where

$$\mathsf{Var}'q_x = \overline{q^2}(x, \omega) - \bar{q}^2(x, \omega)$$

is the variance of the stochastic importance in ω-realization of the random environment. Inserting (53) into (52) yields

$$\mathsf{Var}q_x = \langle \mathsf{Var}'q_x \rangle + \mathsf{Var}\bar{q}(x),$$

where

$$\mathsf{Var}\bar{q}(x) = \langle \bar{q}^2(x) \rangle - \langle \bar{q}(x) \rangle^2 \tag{54}$$

is the variance of semi-stochastic importance, produced by fluctuations of environment. Observe, that in general case, statistical fluctuations of environment change mean value of the stochastic importance and increase its variance.

Observe, that the conditional distributions of the stochastic importance is

$$\Psi(q|x;\omega) = \int \Psi(q|x;\gamma,\omega)P(d\gamma|\omega), \tag{55}$$

whereas the unconditional distribution has the form

$$\Psi(q|x) = \iint \Psi(q|x;\gamma,\omega)P(d\gamma,d\omega). \tag{56}$$

If we are only interested in the influence of the fluctuations of the medium, it is enough to use the distribution of semi-stochastic value

$$\bar{\Psi}(q|x) = \int \delta(q - \bar{q}(x,\omega))P(d\omega).$$

Let us note that the relations (48)–(51) and (54)–(56) lie at the heart of the double-randomization method, which in that before modeling each next trajectory (or a group of a fixed number of trajectories) a random environment realization is modeled.

Modeling a three-dimensional random environment is, for exception maybe a few simple cases, rather a challenging task. This is due, inter alia, to need for the determination of its characteristics in a large number of points of the 3-dimensional region. At the same time, a particle trajectory, unlike, for example, from the wave, occupies only a small part of this space (of zeroth measure): it is a broken line, and its further simulation requires knowledge of the properties only on a set with zeroth measure. The most simple situation is observed in the case of a non-scattering medium: this set is a set of straight lines (rays) of types $\mathbf{r} + x\mathbf{\Omega}$. This allows us to pass from description of a random environment in terms of random fields to the description in terms of random functions of one variable. This is true and for the approximation of "straight-forward", and also, under certain conditions, and for a small-angle approximation.

Below we will consider several models of a random environment, limiting ourselves to study of the non-scattered radiation component, the magnitude of which is determined by the cross-section of the interaction on the ray $\mathbf{r} + x\mathbf{\Omega}$ as a random function x:

$$\Sigma(\mathbf{r} + x\mathbf{\Omega}) = \Sigma(x).$$

6 Gaussian (Normal) Environment

The Gaussian model of a random medium is used in calculations of the passage radiation through the Earth's atmosphere, taking into account its turbulence and other features.

Recall that a random function $\Sigma(x)$ is considered given if for any number n of arbitrarily chosen values $x_1, \ldots, x_n$ the joint distribution

$$w_n(x_1, y_1; \ldots, x_n, y_n) dy_1 \ldots dy_n = \mathsf{P}\{\Sigma(x_1) \in dy_1; \ldots; \Sigma(x_n) \in dy_n\}. \quad (57)$$

is known. All finite-dimensional distributions of w_n are symmetric with respect to permutations of pairs of arguments (x_i, y_i) and, in addition, are consistent between themselves:

$$w_n(x_1, y_1; \ldots, x_k, y_k) = \int dy_{k+1} \ldots \int dy_n w_n(x_1, y_1; \ldots, x_n, y_n), \quad n > k. \quad (58)$$

For Gaussian medium

$$w_n = \frac{1}{\sqrt{(2\pi)^n \Delta}} \exp\left\{-\frac{1}{2\Delta}\Sigma_{i,j}\Delta_{ij}\left(y_i - \langle\Sigma(x_i)\rangle\right)\left(y_j - \langle\Sigma(x_j)\rangle\right)\right\},$$

where $\Delta = \|R_\Sigma(x_i, x_j)\|$ is the n th order determinant compiled by from the values of the correlation function

$$R_\Sigma(x_i, x_j) = \left\langle [\Sigma(x_i) - \langle\Sigma(x_i)\rangle] - [\Sigma(x_j) - \langle\Sigma(x_j)\rangle]\right\rangle$$

Δ_{ij} is an algebraic complement to the corresponding element of determinant Δ. For a statistically homogeneous medium, $\langle\Sigma(x)\rangle$ is independent of x, and $R_\Sigma(x_i, x_j)$ depends only on the modulus of the difference of the arguments

$$x_{ij} = |x_i - x_j|.$$

In this case, the environment is specified by one number $\langle\Sigma\rangle$ and the function of one variable

$$R_\Sigma(x) = \langle\Sigma\rangle^2 \langle\epsilon(0)\epsilon(x)\rangle,$$

where

$$\epsilon(x) = [\Sigma(x) - \langle\Sigma\rangle]/\langle\Sigma\rangle$$

denotes the relative variance of the cross-section Σ. In case of the zero argument, the correlation function coincides with variance, and in a statistically homogeneous medium does not depend on coordinates:

$$R_\Sigma(0) = \langle\Sigma\rangle^2 \langle\epsilon^2\rangle.$$

The correlation function, are often represented in the Gaussian form

$$R_\Sigma(x) = \sigma_\Sigma^2 e^{-x^2/2r^2}$$

and exponential form

$$R_\Sigma(x) = \sigma_\Sigma^2 \Sigma e^{-x/r}.$$

Chaps In both forms, r stands for the correlation radius of the medium, i.e. such a distance, on which the correlation function decreases to about half of its maximum value.

In the general case, the effective correlation radius is given by

$$r_{\text{eff}} = \int\limits_0^\infty R_\Sigma(x)dx / R_\Sigma(0).$$

For the Gaussian form $r_{\text{eff}} = \sqrt{\pi/2}r$, for the exponential $r_{\text{eff}} = r$. However, it may diverge even with disappearing at $x \to \infty$ correlations, as in the case of

$$R_\Sigma(x) = \sigma_\Sigma^2 (1 + x^2/a^2)^{-\alpha}, \quad \alpha \le 1/2.$$

When the correlation radius is much smaller than the average mean free path of the particles in environment, it is convenient to use the white noise model (delta-correlated medium)

$$R_\Sigma(x) = B\delta(x),$$

in which $r_{\text{eff}} = 0$. The constant (in a homogeneous medium)

$$B = 2 \lim r_{\text{eff}} \sigma_\Sigma^2$$

is called the *intensity of white noise*. The most important property of Gaussian random functions is that, the derivatives and definite integrals of them are also distributed normally. It follows that the optical distance between the origin coordinates and the point x

$$\tau(x) = \int_0^x \Sigma(x')dx'$$

has a normal distribution with an average value

$$\langle \tau \rangle = \int\limits_0^x \langle \Sigma(x') \rangle dx' = \langle \Sigma \rangle x$$

and variance

$$\sigma_\tau^2 = \int_0^x dx_1 \int_0^x dx_2 \langle \Sigma(x_1)\Sigma(x_2)\rangle - \langle \tau \rangle^2 = \int_0^x dx_1 \int_0^x dx_2 R_\Sigma(x_{12}).$$

For a random medium with a correlation function in exponential form

$$\sigma_\tau^2 = 2r^2\sigma_\Sigma^2(x/r + e^{-x/r} - 1),$$

and in the case of white noise

$$\sigma_\tau^2 = Bx.$$

If the detector is located in the plane $x = d$ and measures the number of incident particles, then the semi-stochastic importance

$$\bar{q}(x) = e^{-\tau(x,d)}$$

is obviously distributed according to the log-normal law

$$\bar{\Psi}(q|x) = \frac{1}{\sqrt{2\pi}\sigma_\tau q} \exp\left\{-[\ln q + \langle \tau \rangle]^2/(2\sigma_\tau^2)\right\} \tag{59}$$

with the average

$$\langle q \rangle = e^{-\langle \tau \rangle + \sigma_\tau^2/2} \tag{60}$$

and the variance

$$\sigma_q^2 = [e^{\sigma_\tau^2} - 1]\langle q \rangle^2. \tag{61}$$

Observe, that distribution (59) is asymmetric: its mode (the most probable value)

$$\hat{q} = e^{-\langle \tau \rangle} - \sigma_\tau^2 \tag{62}$$

less than the average value (57) and even less than average importance

$$\bar{q}(\langle \tau \rangle) = e^{-\langle \tau \rangle}$$

in a deterministic medium of optical thickness $\langle \tau \rangle$. Inequality

$$\langle q \rangle > \bar{q}(\langle \tau \rangle)$$

It is noted in many work [10], however, inequality

$$\hat{q} < \bar{q}(\langle \tau \rangle)$$

indicates the need for additional research on the tasks to answer the question: what is the best result you are looking for, $\langle \tau \rangle$ or $\hat{q}$?

7 A Homogeneous Medium with Inclusions of Finite Sizes

In this model it is assumed that the physical properties of the realizations media are described by a piecewise constant function that can take two values. One of them characterizes the main phase medium (binder), the other—the phase of the inclusions. We will assume that each of the inclusions is a bounded region, occupied by a homogeneous phase B, and the intervals between inclusions are filled with a homogeneous phase A. Such a model simulates inhomogeneities in the ground, building materials and structures, in boiling liquids etc. To completely define a random environment in this model, one has to describe (in the probabilistic sense) the shape and size of the inclusions, and so is their location in the mainstream.

The simplest model of inclusions is balls. It is easy to see that the length of the chord of a random line intersecting a ball with diameter d, has a distribution density

$$f(x|d) = \begin{cases} 2x/d^2, & x < d, \\ 0, & x > d. \end{cases} \tag{63}$$

If the diameter of the ball is a random variable, it must be the corresponding distribution is given. This may be one of two distributions differing in the procedure for selecting balls from the ensemble. In the first case, it is assumed that from a large number of spheres, placed in this volume, randomly selected one. This density we denote $g_1(y)$. According to the above procedure, $g_1(y)dy$ gives us the fraction of balls in the volume whose diameter belongs to the interval $(y, y + dy)$. In the second case, the choice of balls is carried out with the help of a random straight line drawn in space. The distribution density of diameters of the balls, "punctured" by the straight line, is denoted by $g_2(y)$. Obviously, the densities $g_1(y)$ and $g_2(y)$ do not coincide: in the second case, more often are met the balls of large diameters, since the probability the intersection of the ball is proportional to the area of its principal cross section. $\pi y^2/4$. Consequently,

$$g_2(y) \sim \frac{\pi y^2}{4} g_1(y),$$

or, after renormalization,

$$g_2(y) = \frac{y^2 2\, g_1(y)}{\int\limits_0^\infty y^2 2\, g_1(y)dy}. \tag{64}$$

Thus, if the density $g_2(y)$ is given, then the unconditional distribution of the chord crossing an individual ball has the density

$$f(x) = \int\limits_0^\infty f(x|y)g_2(y)dy = 2x \int\limits_x^\infty y^{-2} g_2(y)dy. \tag{65}$$

But, as it often occurs, if $g_1(y)$ is given, then Eq. (64) may be used, and we get [12]

$$f(x) = 2x \frac{\int_x^\infty g_1(y)dy}{\int_0^\infty y^2 2 g_1(y)dy}. \tag{66}$$

For $g_1(y) = \delta(y - d)$, we return to the original formula (63).

Continuing similar reasoning leads to conclusion that the diameters of inclusions covering a fixed point on a line are distributed according to density

$$g_3(y) = \frac{y^3 g_1(y)}{\int_0^\infty y^3 g_1(y)dy}. \tag{67}$$

Substituting (67) into (65) instead of g_2, we obtain the distribution chord balls of this ensemble. Of great importance is the formula (67) in calculating the probability of the total covering of the segment $[0, l]$ by one ball. This probability $P_1(l)$ can be represented the product of the conditional probability $P_1(l|0)$ of this event (provided that the initial point of the segment 0 is covered, say, on the likelihood of this condition being met (probability covering the point 0, equal to the average volume fraction of inclusions ν)

$$P_1(l) = P_1(l|0)\nu.$$

In its turn

$$P_1(l|0) = \int P_1(l|0, y)g_3(y)dy,$$

$$P_1(l|0, y) = \int_l^y f_1(x|0, y)dx, \tag{68}$$

where $f_1(x|0, y)$ is the distribution density of the length of the ray part, from the interior point of the sphere to its surface in an arbitrary direction, averaged over the coordinates of this points and directions for a fixed diameter of the ball y s:

$$f_1(x|0, y) = \frac{3}{2}\left[\frac{1}{y} - \frac{x^2}{y^3}\right], \qquad x < y. \tag{69}$$

Inserting (69) into (68) yields

$$P_1(l|0, y) \approx 1 - \frac{3}{2}\frac{l}{y} \approx e^{-l/\mu_y}, \qquad l \ll \mu_y, \tag{70}$$

where μ_y is the average length of the chord of a ball y in diameter, found from the distribution (63):

$$\mu_y = \int\limits_0^y xf(x|y)dx = \frac{2}{3}y.$$

Let us return to the distribution of $f(x)$ chords of balls intersecting a line. There are many approximations of this function. In [14], the family types of inclusions are approximated by a two-parameter γ-distribution

$$f(x) = [m(mx/\mu)^{m-1}/\mu\Gamma(m)]e^{-mx/\mu}$$

where μ is the average inclusion chord, and the parameter m is a value, inverse to the square of the relative chord fluctuations

$$m = \mu^2/\sigma^2.$$

When $m = 1$, it turns into an exponential distribution

$$f(x) = e^{-x/\mu}/\mu,$$

which plays an important role in the Markov model of the medium and degenerates into δ-distribution as $\mu \to \infty$.

So far we considered a separate inclusion: one isolated ball. Accounting for the cumulative effect of many inclusions is a more complex problem, the correct solution of which is probably available by means of numerical (statistical) modeling or approximate analytical estimating. For example, the distribution density of the fraction $\theta = x/l$ of a segment length l, occupied by inclusions, may be estimated as

$$P(\theta|l) = P_0\delta(\theta) + P_1\delta(\theta - 1) + [1 - P_0 - P_1]\varphi(\theta|l), \qquad (71)$$

where P_0 is the probability that the whole segment $[0, l]$ will be free from inclusions; P_1 denotes probability of full its coverage by inclusions; $\varphi(\theta|l)$ is the distribution density of θ under condition of partial covering of a segment by inclusions. The following approximations are used in (71) probabilities:

$$P_0 = (1 - v)\exp\{-vl/[\mu(1 - v)]\}, \qquad P_1 = v\exp\{-l/\mu\},$$

Denoting by v the volume fraction of inclusions and by μ the average inclusion chord, one can approximate the function φ by β-distribution:

$$\varphi(\theta/l) = \beta(\theta; wl/\lambda, (1 - w)l/\lambda),$$

where

$$w = (v - P_1)/(1 - P_0 - P_1),$$

$$\lambda = \mu(1 - v) \left\{ 1 + \delta^2 - \frac{vl}{\mu} \frac{P_0}{(1 - P_0)(1 - v)^2} \right\},$$

$\delta^2 = \sigma^2/\mu^2$ is the square of the relative fluctuations of the intersection chord one inclusion. Elementary calculations using distribution (71) and the expressions following it show that

$$\langle \theta \rangle = v, \qquad \sigma_\theta^2 = [P_1(1 - w) + v(w - v) + v\lambda(1 - v)/l]/(1 + \lambda/l).$$

We note that distribution (71) can be approximated by expression

$$P(\theta|l) \approx (1 - vl/\mu)\delta(\theta) + (vl/\mu)\beta[\theta; 1/\delta^2, (l/\mu - 1)/\delta^2,]$$

possessing correct mean and variance. Equation (59) also determines the distribution of the semi-stochastic importance. In a non-scattering medium

$$\bar{q} = \exp\{-[\Sigma_A + \Delta\Sigma\theta]l\}, \qquad \Delta\Sigma = \Sigma_B - \Sigma_A,$$

therefore

$$\bar{\Psi}(q|l) = P\left((|\ln q| - \Sigma_A l)/\Delta\Sigma l|l\right)/(\Delta\Sigma l q).$$

The mean value of this distribution

$$\langle \bar{q} \rangle = \exp(-\Sigma_A l)\left\{ P_0 + P_1 e^{-\Delta\Sigma l} + (1 - P_0 - P_1)\Phi[wl/\lambda, l/\lambda, -\Delta\Sigma l]\right\}, \quad (72)$$

where $\Phi(a, b; x)$ denotes the confluent hypergeometric function [10].

8 Poisson's Model of Point Inclusions

Let us pass to the limiting case of the inclusions model: in small volumes ΔV_i in the neighborhoods of the random points $\mathbf{r}_i$ of the original (not necessarily homogeneous) medium, random density perturbations are produced $\rho_i \to \rho_i' = \rho_i + \delta\rho_i$:

$$\delta\rho_i = \begin{cases} \delta\rho(\mathbf{r}), & \mathbf{r} \in \Delta V_i, \\ 0, & \mathbf{r} \bar{\in} \Delta V_i. \end{cases}$$

Since the density perturbation effect in a small volume is expressed in terms of mass perturbation

$$\delta m_i = \int \delta\rho_i dV = \delta \int_{\Delta V_i} \rho dV,$$

the linear sizes of the volumes are small and can be neglected and the mass perturbations can be regarded as point ones. For the sake of convenience only, we will call
these points *atoms*, without putting any physical meaning into this term

The spatial distribution of random points $\mathbf{r}_i$ is defined as follows. We denote by
$V(R)$ the volume of a part of matter inside a sphere of radius R centered at the origin
(for a homogeneous medium $V(R) = 4/3\pi R^3$). Let $V_1, \ldots, V_n, \ldots$ be the volumes
of disjoint domains inside this sphere. Choosing successively and independently of
each other the random positions $\mathbf{r}_1, \mathbf{r}_2$, etc. of the atoms from the uniform over initial
ball distribution, we arrive at the *polynomial distribution*

$$\mathsf{P}\{n(V_1) = k_1, N(V_2) = k_2, \ldots, N(V_m) = k_m\} = \frac{N(V)!}{k_1!, \ldots, k_m!} \, p_1^{k_1} \ldots p_m^{k_m},$$

where

$$p_i = V_i/V(R), \qquad \sum_{i=1}^{m} k_i = N(V).$$

As $R \to \infty$ and $n = N(V)/V(R) = $ const, it tends to multivariate Poissonian distribution [92]:

$$\mathsf{P}\{N(V_1) = k_1, \ldots, N(V_m) = k_m\} \to \frac{\lambda_1^{k_1}}{k_1!} e^{-\lambda_1} \ldots \frac{\lambda_m^{k_m}}{k_m!} e^{-\lambda_m},$$

where $\lambda_i = nV_i$. Thus, in the limit $R \to \infty$ we obtain an ensemble of random atoms
possessing the following properties:

(1) random numbers of atoms $N(V_i)$ in disjoint domains V_i are mutually independent;
(2) the probability p_k to find k points in a given 3-dimensional domain depends only
on the volume of the region, but not on its shape;
(3) this probability is described by the Poisson distribution

$$\mathsf{P}\{N(V_i) = k\} = \frac{\lambda_i^k}{k!} e^{-\lambda_i}, k = 0, \, 1, \, 2, \, \ldots$$

with mean value $\lambda_i = nV_i$. Such a set of random atoms forms the *homogeneous
Poisson ensemble* [13, 14].

The three properties of the model can be considered as its *axioms* and the model
itself admits extension to non-homogeneous case via introducing a mean density
$\bar{n}(\mathbf{r})$ dependent on coordinates and making replacement

$$\lambda_i \mapsto \int_{V_i} \bar{n}(\mathbf{r}) dV.$$

An important consequence follows these axioms. Let us place a cylindrical tube with cross section σ and length $l \gg \sqrt{\sigma}$ in the Poisson medium, and give to this tube segment an increment δl in its length. According to the above axiom, the joint distribution of random number of atoms N on l and δN on δl is the binomial one. Assume, that some particle (say, for a change, the *quantum*) entered this tube from one end and goes along its axis. Taking the binomial distribution, one can see that probability to find no atoms on l and one atom on δl tends to $e^{-\mu l}\mu\delta l$, as $\delta l \to 0$ and $\langle N(\delta l)\rangle \to \mu\delta l$. We have arrived at result which can be interpreted as the Buger-Lambert law for light attenuation. In case of non-homogeneous medium μ depends on l, and the free path l probability density takes the form

$$
p(l) = \mu(l) \exp\left\{ -\int_0^l \mu(x')dx' \right\}.
\tag{73}
$$

We saw, that the main assumption made in validation of this model, was independence of atoms in their arrangement. Physically, it means the absence of interactions between them. The atoms almost do not feel each other. This takes place in a gaseous medium and is strictly fulfilled in the theoretical model of the *ideal gas*. I'd like to stress, that namely this assumption is the main reason why the kinetic (more correctly, *gas-kinetic*) equations begin with "words"

$$
\frac{\partial f}{\partial t} + \mathbf{v}\frac{\partial f}{\partial \mathbf{r}} - \mu f = \ldots
$$

whereas equation describing transport in a fractal medium begins with factional power of this operator, arising as a result of averaging distribution (73) over the fractal structure.

One should note that in case of a homogeneous point system with long-range correlations of inverse power type (fractal structures), the free path between collisions of a quantum propagating though this system is distributed according inverse power law as well. This fact may seem trivial, but it is not so: exponents in correlation function and in free path distributions are different. Mathematical description of this process can be found in my works [15–20].

9 Inhomogeneous Randomized Poisson Model

One of the most important ways of studying cosmic rays and their interactions with nuclei is measurement of spatial distributions of secondary particles in cascades (called *extensive atmospheric showers* (EAS) triggered by primary particles (protons, gamma quanta) of high energies. Registration of these distributions is carried out by systems of measuring devices placed over surface of the EAS-station. The main cause of inaccuracy of the inferences from results of these measurements is their statistical fluctuations. The peculiarity of the experimental determination of fluctuations in the number of particles arriving at the detector lies in the fact that its dimensions can be directed to zero, as is assumed in the theoretical determination of the flux density, but must remain a non-vanishing parameter of the problem. Direct simulation of a high-energy shower down to the small-size site in question requires a large expenditure of computer time. The partial averaging of showers used in a number of algorithms leads to a distortion of the fluctuations. In particular, the breaking off trajectory of a particle with the replacement of its stochastic importance (that is, the random contribution of its descendants into the detector readings) by a deterministic average value excludes the dependence of the fluctuations on sizes of the detector. The term "density fluctuations" looks here in not quite correct: the random density of the number of point particles is described by a generalized function (the sum of δ-functions containing the coordinates of each particle of the shower at the observation level), whose square, and, consequently, the variance does not exist. In other works, when simulating the experimental situation, the number of particles incident on the detector is modeled according the Poisson distribution. Such an estimate, admitted in the case of a small detector, increases with the detector size that conflicts with fluctuations in the total number of particles in the shower, which are many times larger than the Poisson fluctuations with the same mean value.

Article [21] presented a phenomenological approach to this, based on a special representation of the random distribution of particles over the observation plane. It should satisfy the following demands: (1) fluctuations of small-size detector reading should be close to Poissonian type, (2) fluctuations of a large size detector reading should agree with numerical results, obtained for the total number of particles by spline method [22], and (3) intermediate behavior of fluctuations between these extreme cases should agree with the Monte Carlo simulation results. It turns out that such a model can be constructed on the basis of the simplest assumption about the nature of the particle distribution without involving free parameters not related to the known characteristics of cosmic ray extensive air showers (Fig. 1).

In order to explain the idea, let us divide the observation plane perpendicular to the shower axis into domains a_k ($k = 1, \ldots, m$) and denote by N_k the random number of particles falling on a_k, by $N = \sum_{k=1}^{m} N_k$ the total number of particles in the observation plane, and by $P(n_1, \ldots, n_m | n)$ probability of the event $\{N_1 = n_1, \ldots, N_m = n_m\}$ under condition that $N = n$. The basic assumption of the model: *the conditional distribution is represented by the polynomial law*

$$P(n_1, \ldots, n_m | n) = \frac{n!}{n_1! \ldots n_m!} q_1^{n_1} \ldots q_m^{n_m}, \tag{74}$$

where $q_k > 0$, $\sum_{k=1}^{m} q_k = 1$, $\sum_{k=1}^{m} n_k = n$. In other words, coordinates of each particle on the observation point does not depends on positions of all other particles.

For continuing computation, we pass from probability distribution (74) to generating function

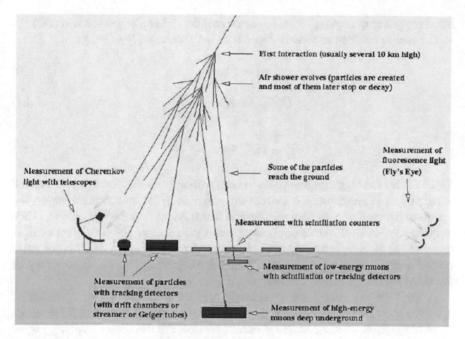

Fig. 1 Measuring cosmic-ray and gamma-ray air showers

$$F(u) \equiv \langle u_1^{N_1} \ldots u_m^{N_m} \rangle = \langle (q_1 u_1 + \ldots + q_m u_m)^N \rangle. \tag{75}$$

The first and second factorial moments of random number N_k are found by differentiation of (75) with respect to u_k at $u_k = 1$:

$$\bar{N}_k = (\partial F / \partial u_k)_{u=1} = \langle N q_k \rangle, \tag{76}$$

$$N_k^{[2]} = (\partial^2 2 \, F / \partial u_k^2)_{u=1} = \langle N(N-1) q_k^2 \rangle.$$

The square of relative fluctuations

$$\delta_{N_k}^2 = \langle N(N-1) q_k^2 \rangle \langle N q_k \rangle^{-2} - 1 + \langle N q_k \rangle^{-1} \tag{77}$$

consists of two parts with different asymptotical behavior as $a_k \to 0$. The first of them,

$$(\delta_{N_k}^2)_{PC} = \langle N(N-1) q_k^2 \rangle \langle N q_k \rangle^{-2} - 1, \tag{78}$$

having a finite limit at $a_k \to 0$, can be called the *regular component* (RC) of fluctuations, whereas the second part

$$(\delta_{N_k}^2)_{PC} = \langle N q_k \rangle^{-1}, \tag{79}$$

infinitely growing as $q_k(a_k) \sim a_k \to 0$, we call the Poissonian component (PC). On assumption that q_k weakly depends on N, formulas (76)–(79) yield

$$q_k = \bar{N}_k/\bar{N}, \qquad (80)$$

$$(\delta^2_{N_k})_{\text{PC}} = \delta^2_N - 1/\bar{N}, \qquad (81)$$

and

$$\delta^2_{N_k} = \delta^2_N - 1/\bar{N} + 1/\bar{N}_k. \qquad (82)$$

This satisfies the above requirements in both extreme cases of large ($\bar{N}_k \to \bar{N}$) and small ($\bar{N}_k \ll \bar{N}$) detectors. To analyze the situation in the intermediate region, the results of Monte Carlo simulations of muon distributions with energies above 5 GeV at the mountain level from the primary proton with energy 10^6 GeV were used. A study of the fluctuations in number of muons arriving at areas of different sizes, placed at different distances from the shower axis, confirmed the applicability of formula (82) for the muon component of EAS.

A similar comparison for the electronic component of the shower could not be achieved due to the lack of results of a complete (analog) simulation of the development of the EAS, up to crossing the small size area. For this reason, another way of testing the model was used, based on comparison with the results of incomplete EAS simulation. For definiteness, let us assume that only the nuclear-active component is modeled, and the random contributions of π^0-decay photons (i.e. their stochastic importance) to the spatial distribution of electrons are replaced by their average values (as is usually done in Monte Carlo calculations. Let $x_1, x_2, \ldots$ be the phase points of the photon production in the random shower realization, and $N(x_i \to a_k)$ be the random number of electrons in the electron-photon cascade generated by the photon coming to the area a_k from x_i. The total number of electrons arriving at a_k in a given shower realization,

$$N_k = \sum_i N(x_i \to a_k). \qquad (83)$$

In case of incomplete simulations, we have instead of (70) another random variable

$$N'_k = \sum_i \bar{N}(x_i \to a_k). \qquad (84)$$

Observe, that the mean values of both variables coincide and fluctuations are linked with relation

$$(\delta^2_{N_k})_{\text{RC}} = \delta^2_{N'_k} + Q/\bar{N}^2_k, \qquad (85)$$

where

$$Q = \int S(x)[\overline{N^2}(x \to a_k) - \bar{N}^2(x \to a_k)]dx,$$

and $S(x)$) is the mean density of photons source in a shower. As $a_k \to 0$ (the area concentrates at a point, position of which is represented by two-dimensional vector **r**),

$$N'_k \approx f(\mathbf{r})a_k,$$

$$\overline{N^2}(x \to a_k) - \bar{N}^2(x \to a_k) \sim \bar{N}(x \to a_k),$$

$$Q \sim \int S(x)\bar{N}(x \to a_k)dx = \bar{N}_k$$

and instead (72) we have

$$\delta^2_{N_k} = \delta^2_{f(\mathbf{r})} + 1/\bar{N}_k \tag{86}$$

This result agrees with the expansion of the fluctuations into two components, the role of the RC fluctuations being fluctuations of the quantity (84) or (which is the same) fluctuations in the quantity

$$f(\mathbf{r}) = \sum_i \bar{f}(x_i \to \mathbf{r}),$$

$$f(x_i \to \mathbf{r}) = \lim_{\sigma_k \to 0} [\bar{N}(x_i \to a_k)/a_k],$$

which is appropriate to be called the RC of a random density of shower particles. Represent it in the form

$$f(\mathbf{r}) = N\rho(\mathbf{r}), \tag{87}$$

where $\rho(\mathbf{r})$ obeys the condition $\int \rho(\mathbf{r})da = 1$ (Tables 1 and 2).

Observe, that Eq. (81) (without the second term negligible compared to the first)

$$(\delta^2_{N_k}) \approx \delta^2_{f(\mathbf{r})} = \delta^2_N \tag{88}$$

corresponds to representation (87), when function $\rho(\mathbf{r})$, describing shape of the spatial distribution, is deterministic and independent of N. With this

$$q_k = \rho(\mathbf{r})a_k, \tag{89}$$

and RC is due only to total number N fluctuations.

Calculations of the fluctuations in the spatial electron distribution in EAS components showed that the relation (88) does not hold for it, and $\delta^2_{f(\mathbf{r})}$ as a function of the distance r can be both more, and less than δ^2_N. Numerical investigation shown that at the see level, the whole energy interval reveals common general pattern: $\delta_f > \delta_N$ at small distances, and $\delta_f < \delta_N$ at large ones ($r \geq 30$).

A similar situation for the mountain level is observed only up to energy 10^6 GeV, at higher energies an excess of δ_f over δ_N is observed in the entire range of the

Table 1 The electron number density $f(r)$ (upper numbers) and RC (lower numbers) in EAS from primary protons (sea level)

r, m	E_p, GeV			
	10^5	10^6	10^7	10^8
10	1.86	$3.0 \cdot 10^1$	$561 \cdot 10^2$	$8.76 \cdot 10^3$
	1.03	0.75	0.56	0.44
20	$3.14 \cdot 10^{-1}$	5.67	$9.07 \cdot 10^1$	$1.36 \cdot 10^3$
	0.91	0.65	0.48	0.38
100	$1.82 \cdot 10^{-2}$	$3.12 \cdot 10^{-1}$	4.70	$6.55 \cdot 10^1$
	0.74	0.52	0.39	0.31
200	$2.47 \cdot 10^{-3}$	$4.03 \cdot 10^{-2}$	$5.74 \cdot 10^{-1}$	7.53
	0.64	0.45	0.33	0.27
400	$2.92 \cdot 10^{-4}$	$4.46 \cdot 10^{-3}$	$5.92 \cdot 10^{-2}$	$7.22 \cdot 10^{-1}$
	0.55	0.38	0.28	0.23
600	$8.11 \cdot 10^{-5}$	$1.19 \cdot 10^{-3}$	$1.51 \cdot 10^{-2}$	$1.75 \cdot 10^{-1}$
	0.50	0.34	0.26	0.22
1000	$1.60 \cdot 10^{-5}$	$2.21 \cdot 10^{-4}$	$2.64 \cdot 10^{-3}$	$2.89 \cdot 10^{-2}$
	0.44	0.30	0.23	0.21
$\bar{N}$	$6.15 \cdot 10^3$	$1.07 \cdot 10^5$	$1.67 \cdot 10^6$	$2.43 \cdot 10^7$
δ_N	0.97	0.66	0.51	0.39

Note The last two lines are the mean and relative fluctuations of the total number of particles at the observation level

Table 2 Same characteristics of EAS, as in Table 1 (mountain level, 690 g/cm²)

r	E_p			
	10^5	10^6	10^7	10^8
10	$1.33 \cdot 10^{-1}$	$1.99 \cdot 10^2$	$2.54 \cdot 10^3$	$2.89 \cdot 10^4$
	0.35	0.24	0.19	0.19
20	1.95	$2.74 \cdot 10^1$	$3.30 \cdot 10^2$	$3.53 \cdot 10^3$
	0.28	0.20	0.18	0.20
100	$8.96 \cdot 10^{-2}$	1.14	$1.25 \cdot 10^1$	$1.23 \cdot 10^2$
	0.21	0.17	0.19	0.22
200	$1.02 \cdot 10^{-2}$	$1.20 \cdot 10^{-1}$	1.23	$1.13 \cdot 10^1$
	0.18	0.17	0.21	0.24
400	$9.79 \cdot 10^{-4}$	$1.05 \cdot 10^{-2}$	$9.92 \cdot 10^{-2}$	$8.51 \cdot 10^{-1}$
	0.17	0.19	0.23	0.27
600	$2.40 \cdot 10^{-4}$	$2.43 \cdot 10^{-3}$	$2.18 \cdot 10^{-2}$	$1.79 \cdot 10^{-1}$
	0.17	0.21	0.25	0.29
1000	$4.03 \cdot 10^{-5}$	$3.78 \cdot 10^{-4}$	$3.18 \cdot 10^{-3}$	$2.47 \cdot 10^{-2}$
	0.19	0.23	0.27	0.31
$\bar{N}$	$3.80 \cdot 10^4$	$5.33 \cdot 10^5$	$6.44 \cdot 10^6$	$6.99 \cdot 10^7$
δ_N	0.31	0.20	0.16	0.12

Table 3 Moments $\langle v^\alpha \rangle^{1/\alpha}$ in EAS from proton with 10^7 GeV (sea level)

α	$\langle v^\alpha \rangle^{1/\alpha}$		α	$\langle v^\alpha \rangle^{1/\alpha}$	
	Numeric. integration	Approximation (81)		Numeric. integration	Approximation (81)
0.2	0.89	0.89	2.0	1.13	1.15
0.6	0.95	0.95	2.6	1.22	1.25
1.2	1.03	1.03	3.0	1.28	1.32
1.6	1.08	1.09	3.6	1.45	1.53

distances under consideration (10–1000 m). The Poisson contribution, depended on the detector area, rapidly increases with increasing distances from the shower axis.

The features noted above point to significant role of *fluctuations in the shape of the distribution* $\rho(\mathbf{r})$ the sign of the correlations between $\rho(\mathbf{r})$ and N with changing the distance from the axis. Taking this effect into account can be realized by replacing the random variables $\rho(\mathbf{r})$ and N with deterministic functions $\rho_0(r, s)$ and $N_0(s)$ of the random parameter s. This representation makes it easy to explain the behavior δ_f with r: a random increase age in N behind the shower maximum means decreasing s, that entails a narrowing $\rho_0(r, s)$ and vice versa, so that the product entails $N_k = N\rho(\mathbf{r})\sigma_k$ fluctuates at small distances in r stronger than N, whereas at large r its fluctuations are smaller than fluctuations in N.

Outside the shower maximum region, the relationship between s and $N_0(s)$ is one-to-one and $\rho_0(r, s)$ can be expressed in terms of N:

$$\rho_0(r, s(N)) = R(r, N). \tag{90}$$

This dependence is easily approximated by a power function

$$R(r, N) = \alpha(r)v^{\gamma(r)}, \qquad v = N/\bar{N}. \tag{91}$$

Using Eqs. (89)–(91), we obtain

$$\bar{N}_k = \langle v^{\gamma+1} \rangle \alpha(r) a_k \bar{N} \tag{92}$$

and

$$(\delta^2_{N_k})_{RC} = \delta^2_{f(r)} \approx \langle v^{2\gamma+2} \rangle \langle v^{\gamma+1} \rangle^{-2} - 1. \tag{93}$$

instead of (76) and (78). On assumption about lognormality of total number of electrons in EAS,

$$\langle v^\alpha \rangle = (1 + \delta^2_N)^{\alpha(\alpha-1)/2}. \tag{94}$$

Table 3 shows that this approximation is in satisfactory agreement with the results of direct numerical integration of the distributions over N obtained by the Monte

Carlo method with high statistics (10^6 showers). From (79)–(81), we find simple formulas for the average number density of particles and RCF for the electrons:

$$\bar{f}(r) = \alpha(r)\bar{N}(1 + \delta_N^2)^{(\gamma+1)\gamma/2}, \tag{95}$$

$$\delta_{f(r)}^2 = (1 + \delta_N^2)^{(1+\gamma)^2} - 1. \tag{96}$$

The coefficients α and γ can be determined by comparison with the results of numerical calculations. In this case, formula (96) is simply a more or less convenient approximation of the results of numerical calculations. But formulas (95, 96) can be used for other purposes: making sure that $\alpha(r)$ is close $\bar{\rho}(r) = R(r, \bar{N})$, the approximate values of the coefficient γ can be determined from the formula (91):

$$\tilde{\gamma} = \lg[R(r, N)/R(r, \bar{N})]/\lg[N/\bar{N}],$$

$$N/\bar{N} \in (1 - \delta_N, 1 + \delta_N). \tag{97}$$

In this case, formula (96) makes it possible to obtain approximate fluctuations $\tilde{\delta}_{f(r)}$ and without numerical calculations, relying only on the known average characteristics of showers and fluctuations of the total number of particles. Checking shows the satisfactory agreement of $\tilde{\delta}_{f(r)}$ with $\delta_{f(r)}$ in the region where the PC is not predominant. This is an important argument in favor of the phenomenological approach developed in the present paper, as a simpler and less elementary process of the multiple process that is less connected with the particular features of this or that model. As the level of observation increases, we approach the highs cascade curves in the neighborhood of which the connection between s and N ceases to be one-to-one and $\rho_0(r, s)$ can not be approximated by formula (91). For this reason, formula (92) does not give correct results for the level of mountains. The calculation of the fluctuations in this case can also be made involving additional calculations (see [21]) and using Eq. (86). In any case, the decomposition of the fluctuations into two noncorrelating components, RC and PC, called the *two-component model* s useful for processing EAS measurement data the fluctuations in the spatial distribution of shower particles.

10   Field, Generated by Poissonian Point Ensemble

Here we are going to discuss a continuous version of random environment: the field created by a Poisson system of point sources. this model comes from the classic work on the statistics of stars and the fields (light and gravitational) they create in the Galaxy (Holtsmark, 1919), used by Chandrasekhar and von Neumann (Chandrasekhar & Neumann, 1941; Chandrasekhar & Neumann, 1943; Chandrasekhar, 1944a; Chandrasekhar, 1944b), and then generalized by Zolotarev (Zolotarev, 1986), and Uchaikin and Zolotarev (Uchaikin and Zolotarev, 1999) [13], [24]–[29].

We consider a simplified version of this model with 3-dimensional, possible bounded in space Poisson ensemble of point sources with intensities $q_i, i = 1, 2, 3, \ldots$, located at random points $\mathbf{r_i}$. Let us call them *charges*, in order to allow them to take both positive and negative values. At the origin, which we chosen as the observation point, this ensemble creates a scalar field

$$\Phi(0) = \sum_i \varphi_i(\mathbf{r}_i). \tag{98}$$

called *potential*, and each of charges contributes to this potential independently of all others the quantity

$$\varphi_i(\mathbf{r}_i) = q_i g(\mathbf{r}_i).$$

Assume, that all q_i are mutually independent and independent of $\mathbf{r}_i$ and

$$g(\mathbf{r}) \sim r^{-2} A(\mathbf{r}/r), \quad r = |\mathbf{r}| \to 0, \tag{99}$$

Let us introduce designation V_x for the domain, where $g(\mathbf{r}) > x$, if $x > 0$, and $g(\mathbf{r}) < x$, if $x < 0$:

$$V_x = \begin{cases} \int\limits_{g>x} dV, x > 0, \\ \int\limits_{g<x} dV, x < 0. \end{cases}$$

Inserting here the asymptotical expression (99), one can estimate V_x at large in absolute x:

$$V_x \sim \begin{cases} \int\limits_{\Omega_+} d\Omega \int\limits_0^{\sqrt{A/x}} r^2 dr = a_1 x^{-3/2}, \quad x \to \infty, \\ \int\limits_{\Omega_-} d\Omega \int\limits_0^{\sqrt{|A|/x}} r^2 dr = a_1 |x|^{-3/2}, x \to -\infty, \end{cases}$$

where

$$\Omega_+ = \{\mathbf{\Omega} : A(\mathbf{\Omega}) > 0\}, \qquad \Omega_- = \{\mathbf{\Omega} : A(\mathbf{\Omega}) < 0\},$$

$$a_1 = \frac{1}{3} \int\limits_{\Omega_+} [A(\mathbf{\Omega})]^{3/2} d\mathbf{\Omega}, \qquad b_1 = \frac{1}{3} \int\limits_{\Omega_-} |A(\mathbf{\Omega})|^{3/2} d\mathbf{\Omega},$$

Returning to the sum of the random variables (59), we put first $q_i = \text{const} = 1$, then (59) takes the form

$$\Phi(0) = \sum_{i=1}^n \xi_i \equiv S_n,$$

where ξ_i are independent random summands with the same distribution law that has asymptotics

$$P\{\xi > x\} = P\{\mathbf{r}_i \in V_x\} = V_x/V(R) \sim [a_1/V(R)]x^{-3/2}, \qquad x \to \infty;$$

$$P\{\xi < -|x|\} = P\{\mathbf{r}_i \in V_{-|x|}\} = V_{-|x|}/V(R) \sim [b_1/V(R)]|x|^{-3/2}, \quad x \to -\infty.$$

According to the Levy-stable laws theory

$$P\left\{\frac{S_n - \langle S_n \rangle}{[nc(a_1 + b_1)]^{2/3}} < x\right\} \Rightarrow F_{3/2,\beta}(x), \quad n \to \infty, \tag{100}$$

where $F_{3/2,\beta}$ is the cumulative stable distribution function with characteristic exponent $\alpha = 3/2$, $\beta = (a_1 - b_1)/(a_1 + b_1)$ is skewness parameters, c is a scale constant.

For random q_i, positive with probability p_+ and negative with probability p_-, which have finite conditional moments

$$\langle q^{3/2} \rangle_+ = \langle q^{3/2} \rangle |_{q>0},$$

$$\langle |q|^{3/2} \rangle_- = \langle |q|^{3/2} \rangle |_{q<0},$$

conclusion (100) remains valid after replacement $a_1 \to a$ and $b_1 \to b$, where

$$a = \frac{1}{3}\left\{ p_+\langle q^{3/2}\rangle + \int_{\Omega_+} A^{3/2}(\mathbf{\Omega})d\mathbf{\Omega} \right\},$$

$$b = \frac{1}{3}\left\{ p_-\langle |q|^{3/2}\rangle - \int_{\Omega_-} |A(\mathbf{\Omega})|^{3/2}d\mathbf{\Omega} \right\},$$

When $\beta = 0$, the $\langle S_n \rangle \to 0$, and the limit random variable

$$\frac{S_n}{[nc(a_1 + b_1)]^{2/3}} \sim Y^{(3/2)}, \quad n \to \infty,$$

has the symmetric distribution with characteristic function

$$\mathsf{E}\, e^{ikY^{(3/2)}} = e^{-|k|^{3/2}}.$$

is analogous to the solutions of Holtzmark problems and Olbers in astrophysics (fluctuations of gravitational forces and light flux in a Poisson ensemble of stars). In particular, the random gravitational field strength $\mathbf{F}$ created by Poissonian ensemble os stars has a characteristic function

$$\mathsf{E}\, e^{i\mathbf{k}\cdot\mathbf{F}} = e^{-|\mathbf{k}|^{3/2}},$$

looking like an obvious generalization of the above formula. Three-dimensional distributions related to characteristic functions $e^{-|\mathbf{k}|^{\alpha}}$ with $\alpha \in (0, 2]$ form the isotropic Levy-Feldheim family. The inverse Fourier transform yields

$$f_{\alpha}(\mathbf{r}) = (2\pi)^{-3} \int\limits_{\mathbb{R}^3} e^{-i\mathbf{k}\cdot\mathbf{r}-|\mathbf{k}|^{\alpha}} d\mathbf{k} = \frac{1}{2\pi^2 r} \int\limits_{0}^{\infty} e^{-k^{\alpha}} \sin(kr)k dk.$$

Case $\alpha = 3/2$ corresponds to the Holtzmark distribution, cases $\alpha = 1$ and $\alpha = 2$ admit integration in elementary functions, leading to three-dimensional Cauchy's and Gauss' distribution (with doubled variance) respectively:

$$f_1(\mathbf{r}) = \frac{1}{\pi^2(1+r^2)^2}, \quad f_2(\mathbf{r}) = \frac{1}{(4\pi)^{3/2}} e^{-r^2/4}.$$

At the end of this section, I would like to draw the reader's attention to the words spoken at the beginning of the chapter: *within the framework of the linear approximation of perturbation theory, the influence of a set of local inhomogeneities on the detector reading is equivalent to a system of additional local sources. placed in homogeneous medium.* This is the main purpose of this section, and in particular, This is the main purpose of this section, and in this, in particular, the reason why we admit the sources to be anisotropic.

11   Stratified Random Environment of Markov Type

We now turn to the consideration of models of piecewise homogeneous media, limiting ourselves here to the case of a stratified (stratified) medium Let us return to description of random point distribution on the axis $-\infty < x < \infty$ and represent the n-variable density in the form

$$w_n(x_1, y_1; \dots; x_n, y_n) = w_{n-1}(x_1, y_1; \dots; x_{n-1}, y_{n-1}) \times$$

$$\times\ v_n(x_n, y_n | x_1, y_1; \dots; x_{n-1}, y_{n-1}). \tag{101}$$

The factor v_n on the right-hand side is the conditional probability density of the random variable $\Sigma(x)$ under the condition that at the previous points of the segment $x_1, \dots, x_{n-1}$ the function $\Sigma(x)$ took values $y_1, \dots, y_{n-1}$. If for any choice of x_i and a fixed y_{n-1}, the conditional density does not depend on $y_{n-2}, y_{n-3}, \dots, y_1$ described by the distribution (101) A random process is called it Markov (the variable x is interpreted as time), and $\Sigma(x)$ is a random function Markov type. Thus, if it is known that in some "time moment" x_{n-1} is a random Markov type function $\Sigma(x)$ had the value y_{n-1}, then its further behavior (in the probabilistic sense) is uniquely defined and does not depend on any information at all about its prehistory (to the moment x_{n-1})

$$v_n(x_n, y_n|x_1, y_1; \ldots, x_{n-1}, y_{n_1}) = v(x_n, y_n|x_{n-1}, y_{n-1}). \tag{102}$$

From (101) and (102) follows, that

$$w_n(x_1, y_1; \ldots, x_n, y_n) = w_{n-1}(x_1, y_1; \ldots, x_{n-1}, y_{n-1})v(x_n, y_n|x_{n-1}, y_{n-1}) \tag{103}$$

or

$$w_n(x_1, y_1; \ldots, x_n, y_n) = w_1(x_1, y_1)v(x_2, y_2|x_1, y_1) \ldots v(x_n, y_n|x_{n-1}, y_{n-1}). \tag{104}$$

Therefore, in order to obtain the n-dimensional probability density of a Markov process for $x \geqslant x_1$, you only need to know two functions—one-dimensional density probability w_1 at the starting point x_1 and the conditional density $v(x, y|x', y')$ for all $x, x' > x_1$. The latter is often called the *transition density*. A random process (a random function) is called *homogeneous*, if the transition density $v(x, y|x', y')$ as a function of two arguments x, x' depends on their difference, and not on each of them separately:

$$v(x, y|x', y') = v(y|x - x', y').$$

If $t \to 0$ limits

$$\lim_{t \to 0} \frac{1}{t} \int_{-\infty}^{\infty} (y - y')v(x, y|x - t, y')dy = A(y', t),$$

$$\lim_{t \to 0} \frac{1}{t} \int_{-\infty}^{\infty} (y - y')^2 v(x, y|x - t, y')dy = B(y', t),$$

$$\lim_{t \to 0} \frac{1}{t} \int_{-\infty}^{\infty} |y - y'|^3 v(x, y|x - t, y')dy = 0,$$

exist, we deal with a diffusion process, described by diffusion equation and leading, in particular, to Gaussian distributions. In case

$$v(y|t, y') \sim [1 - \sigma(y)t]\delta(y - y') + t\sigma(y' \to y) \text{ as } t \to 0, \tag{105}$$

the Markovian process becomes jump-process. Integrating both sides of (103) with respect to $y_1, \ldots, y_{n-1}$ and taking into account self-consistency condition (47), after replacement $y_n = y$, $y_{n-1} = y'$, $x_n = x$, $x_{n-1} = x - t$ and substitution (105) we obtain a differential interrelation for probability density $p(x, y) \equiv w_{x,y}$

$$p(x, y) = [1 - \sigma(y)t]p(x - t, y) + t \int dy' p(x - t, y)\sigma(y' \to y),$$

leading to integro-differential equation

$$\partial p/\partial x + \sigma p = \int dy' p(x, y')\sigma(y' \to y). \tag{106}$$

Here, the integration on the right-hand side is performed over the whole range of values random function Σ. If the set of its values is discrete, then instead of the probability density one should use the probability $p_k(x)$ that $\Sigma(x) = \Sigma_k$. Instead of (106), we have in this case

$$\partial p_k/\partial x + \sigma_k p_k = \sum_{j \neq k} p_j \sigma_{j \to k}. \tag{107}$$

Both these equations should be added by initial conditions, giving $p(0, y)$ in the first case and $p_k(0)$ in the second.

Equation (107) form the basis of the random environment model, called the *Markov model* [31]–[36]. In its framework it is assumed that the physical properties of the medium's realizations along the x-axes are described by piecewise constant functions the set of their values is finite. Constancy of property in a given interval Δx means that it is filled with a homogeneous substance of a certain type. We call these substances *phases*. Single-phase medium is a homogeneous medium, two-phase medium is the simplest version of this models, consisting of alternating segments of both phases of different lengths: if there is a phase 1 to the left of the boundary, then to the right should be 2, and on the contrary. In a multiphase medium to the right of the right boundary of phase 1, phase of any type $i \neq 1$. Different realizations of a random multiphase medium differ from one another in boundaries and the sequence of phases filling the gaps between them. Probabilistic measure in the space of these realizations and set by means of a Markov jump process with a continuous time, whose role is played by the spatial coordinate x. For this it is sufficient to set the initial probabilities $p_k(0)$ and probabilities transitions per unit time $\sigma_{j \to k}$, and

$$\sigma_j = \sum_{k \neq j} \sigma_{j \to k}.$$

The solution of system (107) can be represented in the form

$$p_k(x) = \sum_i p_i(0) p_{i \to k}(x), \tag{108}$$

where $p_{i \to k}$ are transition $(i \to k)$ probabilities on a segment of length x, satisfying equation (107) with special initial conditions:

$$\partial p_{i\to k}/\partial x + \sigma_k p_{i\to k} = \sum_{j\neq k} p_{i\to j}\sigma_{j\to k}, \quad p_{i\to k}(0) = \delta_{ik}. \tag{109}$$

In particular, for a two-phase random medium

$$\partial p_{1\to 1}/\partial x + \sigma_1 p_{1\to 1} = \sigma_2 p_{1\to 2}, \quad p_{1\to 1}(0) = 1,$$

$$\partial p_{1\to 2}/\partial x + \sigma_2 p_{1\to 2} = \sigma_1 p_{1\to 1}, \quad p_{1\to 2}(0) = 0,$$

$$\partial p_{2\to 1}/\partial x + \sigma_1 p_{2\to 1} = \sigma_2 p_{2\to 2}, \quad p_{2\to 1}(0) = 0,$$

$$\partial p_{2\to 2}/\partial x + \sigma_2 p_{2\to 2} = \sigma_1 p_{2\to 1}, \quad p_{2\to 2}(0) = 1.$$

Solution of this system is of the form

$$p_{1\to 1} = [\lambda_1 + \lambda_2 \exp(-x/\lambda)]/(\lambda_1 + \lambda_2);$$

$$p_{1\to 2} = \lambda_2[1 - \exp(-x/\lambda)]/(\lambda_1 + \lambda_2);$$

$$p_{2\to 1} = \lambda_1[1 - \exp(-x/\lambda)]/(\lambda_1 + \lambda_2);$$

$$p_{2\to 2} = [\lambda_2 + \lambda_1 \exp(-x/\lambda)]/(\lambda_1 + \lambda_2),$$

where

$$\lambda_i = 1/\sigma_i, \quad \lambda = 1/\sigma, \quad \sigma = \sigma_1 + \sigma_2.$$

It is clear, that

$$p_{1\to 1} + p_{1\to 2} = p_{2\to 2} + p_{2\to 1},$$

i.e. only two from four probabilities $p_{i\to j}$ are independent. Using these solutions in Eq. (108) yields, in particular, that

$$p_1(x) = (\lambda_1 + \lambda_2)^{-1}\left\{\lambda_1 + [\lambda_2 p_1(0) - \lambda_1 p_2(0)]\exp(-x/\lambda)\right\}. \tag{110}$$

If

$$\lambda_2 p_1(0) - \lambda_1 p_2(0) = 0,$$

we have constant solutions

$$p_i(x) = p_i(0) = \lambda_i/(\lambda_1 + \lambda_2).$$

For such a statistically homogeneous medium, mean cross-section

$$\langle \Sigma \rangle = p_1 \Sigma_1 + p_2 \Sigma_2$$

and its variance

$$\mathsf{Var}\Sigma = p_1 p_2 (\Sigma_1 - \Sigma_2)^2$$

do not depend on x, but correlation function

$$R_\Sigma(x', x'') = \mathsf{Var}\Sigma \exp(-|x'' - x'|/\lambda), \quad \lambda = (\sigma_1 + \sigma_2)^{-1}, \tag{111}$$

depends only on the distance between points x' and x'' [31].

12 Phason Interpretation of Markovian Medium

Equation (107) allow us to give a new useful interpretation of the Markov random environment, based on their external resemblance to equations, describing transfer of particles of different types moving in the direction of the x-axis. In this interpretation, $p_i(0)$ means the probability of birth a particle of the type i at the point $x = 0$; $p_k(x)$ is the probability that a particle, passing through the point x, belongs to the type k; $\sigma_k(x)dx$ is the probability The fact that a particle of type k on an elementary path dx will change its type (a kind of interaction); $\sigma_{j\rightarrow k}(x)dx$ is the probability a change of type j to a type k of a particle along this path. In this case, the free run a particle of the ith type is associated with the layer of the medium occupied by the i th phase, and the interaction points at the beginning and end of the run with boundaries of i th phase. We call these imaginary particles *phason*. Each such trajectory relates to some realization of the random medium, and their ensemble represents the medium as a whole. Averaging over the realizations is adequate to averaging over phason trajectories, which, in fact, is the meaning of this interpretation.

Multiplying (107) by a large number N, we rewrite the equation in the form

$$\partial \varphi_k / \partial x + \sigma_k \varphi_k = \sum_{j \neq k} \varphi_j \sigma_{j\rightarrow k}, \quad \varphi_k(0) = N p_k(0), \tag{112}$$

where $\varphi_k(x) = N p_k(x)$ can be interpreted as k-phason concentration (or flux) at depth x, and $N = \sum_k \varphi_k(x)$ as the total their concentration at this depth. The mean value of cross-section of interaction of real particles (quanta) with this medium is given by formula

$$\langle \Sigma(x) \rangle = \sum_k \Sigma_k p_k(x) = \frac{\sum_k \Sigma_k \varphi_k(x)}{\sum_k \varphi_k(x)}.$$

The phase interpretation of the Markov environment opens up possibilities for using concepts of importance in relation to the phason. Thus, the random optical thickness of the layer (x, l) starting in the phase i.

$$T_i(x, l) = \int\limits_x^l \Sigma(x')dx',$$

can be regarded as a *linear stochastic importance of the phason i* at the point x. Introducing the distribution functions for it,

$$\Psi_i(\tau|x) = \mathbf{P}(T_i < \tau|x), \quad \psi_i(\tau|x) = \frac{\partial \Psi_i(\tau|x)}{\partial \tau},$$

one can derive the following equation for $\psi_i(\tau|x)$:

$$-\frac{\partial \psi_i}{\partial x} + \Sigma_i \frac{\partial \psi_i}{\partial \tau} + \sigma_i \psi_i = \sum_{j \neq i} \sigma_{i \to j} \psi_j(\tau|x), \quad \psi_i(\tau|l) = \delta(\tau). \tag{113}$$

Simple manipulations with this equation lead to equations for expected value of T

$$-\frac{\partial \langle T \rangle_i}{\partial x} + \sigma_i \langle T \rangle_i = \sum_{j \neq i} \sigma_{i \to j} \langle T \rangle_j + \Sigma_i$$

and its higher moments

$$-\frac{\partial \langle T^n \rangle_i}{\partial x} + \sigma_i \langle T^n \rangle_i = \sum_{j \neq i} \sigma_{i \to j} \langle T^n \rangle_j + n \Sigma_i \langle T^{n-1} \rangle_i.$$

Let us consider in more detail the first of them for the optical thickness of a two-phase medium:

$$-\frac{\partial \langle T \rangle_1}{\partial x} + \sigma_1 [\langle T \rangle_1 - \langle T \rangle_2] = \Sigma_1,$$

$$-\frac{\partial \langle T \rangle_2}{\partial x} + \sigma_2 [\langle T \rangle_2 - \langle T \rangle_1] = \Sigma_2,$$

Performing Laplace transformation we find

$$\langle T \rangle_1 = \frac{\Sigma_1 \sigma_2 + \Sigma_2 \sigma_1}{\sigma_1 + \sigma_2} s + \frac{\sigma_1 (\Sigma_1 - \Sigma_2)}{(\sigma_1 + \sigma_2)^2} [1 - e^{-\sigma s}], \tag{114}$$

where $s = l - x$ is the geometric thickness of the layer. Notice, that at small s

$$\langle T \rangle_1 \sim \Sigma_1 s,$$

whereas at large ones

$$\langle T \rangle_1 \sim_1 (\Sigma_1 \sigma_2 + \Sigma_2 \sigma_1) s / \sigma.$$

Since the expression for $\langle T \rangle_2$ is obtained from (114) by permutation indices, we see that the asymptotics of $\langle \tau \rangle_1$ with $s \to \infty$ is not depended on the initial index. The optical thickness averaged over initial states,

$$\langle T \rangle = (\sigma_2/\sigma)\langle T \rangle_1 + (\sigma_1/\sigma)\langle T \rangle_2 = (\Sigma_1\sigma_2 + \Sigma_2\sigma_1)s/\sigma,$$

so that

$$\langle T \rangle_i \sim \langle T \rangle, \qquad s \to \infty.$$

Equation (113) for the probability densities $\psi_i(\tau|x)$ also have exact analytic solution [31], which leads to the following formula for cumulative distribution function $\Psi(\tau|s)$ for $\Sigma_1 > \Sigma_2$:

$$\Psi(\tau|s) = p_1 \int_0^\tau \psi_1(\tau'|s)d\tau' + p_2 \int_0^\tau \psi_2(\tau'|s)d\tau' =$$

$$= p_1 \left\{ 1 - e^u \left[1 + 2 \int_0^{\sqrt{uv}} dx I_1(2x) e^{-x^2/u} \right] \right\} + \tag{115}$$

$$+ p_2 e^{-v} \left[1 + 2 \int_0^{\sqrt{uv}} dx I_1(2x) e^{-x^2/v} \right], \qquad \Sigma_2 s < \tau < \Sigma_1 s,$$

where I_1 is a modified Bessel function of the first kind,

$$u = \sigma_1 \frac{\tau - \Sigma_2 s}{\Sigma_1 - \Sigma_2}, \qquad v = \sigma_2 \frac{\Sigma_1 s - \tau}{\Sigma_1 - \Sigma_2}.$$

The semistochastic importance of real information carriers

$$\bar{q}_i(x) = e^{-T_i(x,l)} \tag{116}$$

can be considered as a nonlinear (multiplicative) importance of phason . Its average value obeys the equation

$$-\partial \langle q \rangle_i / \partial x + (\sigma_i + \Sigma_i)\langle q \rangle_i = \sum_{j \neq i} \sigma_{i \to j} \langle q \rangle_j, \quad \langle q(l) \rangle_i = 1, \tag{117}$$

which due to lack of multiplication processes remains linear. For the Laplace transform we have

$$(\lambda + \sigma_i + \Sigma_i)\tilde{q}_i = \sum_j \sigma_{i \to j} \tilde{q}_j + 1.$$

In the case of a two-phase medium

$$\tilde{q}_1 = \frac{\lambda + \Sigma_2 + \sigma}{(\lambda - \lambda_1)(\lambda - \lambda_2)},$$

$$\lambda_{1,2} = \frac{1}{2}\left\{-\Sigma - \sigma \pm \sqrt{(\Delta\Sigma + \Delta\sigma)^2 + 4\sigma_1\sigma_2}\right\},$$

$$\Sigma = \Sigma_1 + \Sigma_2, \quad \sigma = \sigma_1 + \sigma_2,$$

$$\Delta\Sigma = \Delta\Sigma_2 - \Delta\Sigma_1, \quad \Delta\sigma = \Delta\sigma_2 - \Delta\sigma_1.$$

Inversion yields

$$\langle q \rangle_1 = \frac{\lambda_1 + \Sigma_2 + \sigma}{\lambda_1 - \lambda_2}e^{\lambda_1 s} + \frac{\lambda_2 + \Sigma_2 + \sigma}{\lambda_2 - \lambda_1}e^{\lambda_2 s}.$$

The second solution $\langle q \rangle_2$ is obtained by permuting the indices. After averaging over the initial states of the phason

$$\langle q \rangle = (\sigma_2/\sigma)\langle q \rangle_1 + (\sigma_1/\sigma)\langle q \rangle_2$$

we arrive at a result that coincides with formula (118) [31] obtained by direct averaging (116) with the use of (115).

13   Stratified Random Medium of the Renewal Type

Let us consider an example of random medium, given in [30]. "It is assumed, that the realizations of the random function $\Sigma(x)$) are constructed as follows. On the x-axis, a Poisson sequence of points $x_0, x_1, x_2, \ldots$ is constructed, that the random variables $\eta_i = x_i - x_{i-1}, i = 1, 2, \ldots$, are independent and distributed with the density $\sigma e^{-\sigma t}$ $(t > 0)$; for each of the interval (x_{i-1}, x_i) we choose our independent value σ from the given dimensional distribution of this quantity. Thus, the constructed random function $\Sigma(x)$ is homogeneous and has a given one-dimensional distribution, and the corresponding the normalized correlation function is equal to $e^{-\sigma x}$. This model is Markovian, since a jump through a given level x is also exponential in distribution; this seems naturally for media including random conglomerates of any inhomogeneities, packed not too closely".

In accordance with the above description of such models, we insert

$$\sigma_i = \sigma = \text{const},$$

in equation (117) and after replacing $\sigma_{i \to j}$ by $\sigma \cdot f(\omega)$ with condition $\int f(\omega)d\omega = 1$. As a result, we will get

$$-\partial\langle q \rangle_\omega / \partial x + (\sigma + \Sigma(\omega))\langle q \rangle_\omega = \sigma \int f(\omega)\langle q \rangle_\omega d\omega,$$

or after Laplace transform

$$[\lambda + \sigma + \Sigma(\omega)]\tilde{q}_\omega = \sigma \int f(\omega)\tilde{q}_\omega d\omega + 1.$$

Averaging this expression in ω and introducing notation

$$\chi(\lambda) = \int f(\omega)\tilde{q}_\omega d\omega, \tag{118}$$

we arrive at relation

$$\chi(\lambda) = [\sigma\chi(\lambda) + 1]\chi_1(\lambda), \tag{119}$$

where

$$\chi_1(\lambda) = \int \frac{f(\omega)d\omega}{\lambda + \sigma + \Sigma(\omega)}.$$

According to (119)
the inverse transform is written as

$$\chi(s) = \frac{1}{2\pi i} \int e^{\lambda s}\chi(\lambda)d\lambda \sim \frac{\chi_1(\lambda_1)}{\sigma\chi_2(\lambda_1)}e^{-\lambda_1 s}, \quad s \to \infty, \tag{120}$$

where λ_1 is supposed to be an extreme right pole of the first-order integrand ($\lambda_1 < 0$), and

$$\chi_2(\lambda) = d\chi_1/d\lambda = \int \frac{f(\omega)d\omega}{[\lambda + \sigma + \Sigma(\omega)]^2}.$$

In view of (116) and the commutation of the operations of integration with respect to λ and ω, expression (120) is a function e^{-T} averaged over all trajectories of the phason, including various initial states, so that

$$\langle q \rangle = \langle e^{-\tau} \rangle = \chi(s).$$

In conclusion, let us return to the system of equation (117) and transform them into integral ones:

$$\langle q(x) \rangle_i = e^{-[\sigma_i + \Sigma_i](l-x)} + \int_x^l dx' e^{-[\sigma_i + \Sigma_i](x'-x)} \sum_{j \neq i} \sigma_i \pi_{i \to j}\langle q(x') \rangle_j. \tag{121}$$

It easily to understand, that

$$e^{-\sigma_i(l-x)} = Q_i(x,l) \tag{122}$$

is the probability that i-phason, being at point x, will pass the segment (x,l) without changing its type. With this probability, Eq. (121) are represented as

$$\langle q(x) \rangle_i = Q_i(x,l)e^{-\Sigma_i(l-x)} - \int_x^l d' Q_i(x,x')e^{-\Sigma_i(x'-x)} \sum_{j \neq i} \pi_{i \to j} \langle q(x') \rangle_j. \tag{123}$$

Equation (123) refer to the type of *renewal equations*, and modeled on their basis media also called *media of renewal type*. In this case, they are a consequence of the Markov property of the medium, but in the general case the system takes the form[1]

$$\langle q(x) \rangle_i = K_i(x,l) - \sum_{j \neq i} \int_x^l d' K_{i \to j}(x,x') \langle q(x') \rangle_j. \tag{124}$$

Different from the Markov ones in that the main property of the latter is independence the future from the past with a fixed present—is not preserved for everyone, but only for certain points x, called recovery points. Such the points in this case are the coordinates of the phase boundaries, so under x in (123) should be understood as the coordinate of the first boundary, the distribution of which $w_i(x)$ to the interval $[0, l]$ must be given separately. Together with the initial probabilities $p_i(0)$ and the solutions of system (123) they give the required value

$$\langle q \rangle = \sum_i p_i(0) \left\{ \int_0^l w_i(x) \sum_{j \neq i} \pi_{i \to j} e^{-\Sigma_i x} \langle q(x) \rangle_2 dx + \left[1 - \int_0^l w_i(x) dx \right] \bar{q}_i(0) \right\},$$

where $\bar{q}_i(x)$ is the mean importance of a particle in the determined homogeneous i-medium.

Other applications of renewal equations to modeling of random media can be found in [31]–[38].

References

1. Marchuk GI, Orlov VV (1961) To the adjoint function theory. In: Neutron physics. Moscow, Gosatomizdat, pp 30–45 (in Russian)
2. Lewins J (1965) Importance, the adjoint function. The physical basis of variational and perturbation theory in transport and diffusion problems. Elsevier Science & Technology

---

[1] Symbol d' in two last equations means the differential with respect to x.

3. Ussachoff LN (1955) Equation for the importance of neutrons, reactor kinetics and the theory of perturbations. In: International conference on the peaceful uses of atomic energy. Geneva
4. Marchuk GI (1975) Methods of numerical mathematics. Springer, New York etc
5. Weinberg AM, Wigner EP (1958) The physical theory of neutron chain reactors. In: The University of Chicago Press, Chicago
6. Kadomtsev BB (1957) On the influence function in transport theory. Doklady Akad Nauk SSSR 113:541–543 (in Russian)
7. Tatarsky VI (1967) Wave propagation in turbulent atmosphere. Nauka, Moscow
8. Rytov SM, Kravtsov YuA, Tatarskii VA (1987) Principles of statistical radiophysics. Springer, Berlin, Heidelberg
9. Shimaru A (1978) Wave propagation and scattering in random media. Academic, New York
10. Williams MMR (1974) Random processes in nuclear reactors. Pergamon Press, N.Y
11. Lappa AV, Uchaikin VV, Kolchuzhkin AM, Suslik AZ (1976) Cross-sections randomization in calculation of particles propagation through inhomogeneous medium. Statistical modeling in mathematical physics, Novosibirsk, Siberian Branch of Academy of Sciences USSR, pp 17–26
12. Kendall MG, Moran PAP (1963) Geometrical probability. Griffin, London
13. Zolotarev VM (1986) One-dimentional stable distributions, translations of mathematical monographs, vol 65. American Mathematical Society, Providence
14. Feller W (1966) An introduction to probability theory and its applications. Willey, New York
15. Uchaikin VV, Gusarov GG (1997) Levy-flight applied to random media problem. J Math Phys 38(5):2453–2464
16. Uchaikin VV, Gismjatov I, Gusarov GG et al (1998) Paired Levy-Mandelbrot trajectory as a homogeneous fractal. Int J Bif Chaos 8(5):977–984
17. Uchaikin VV (2000) Small-angle multiple scattering on a fractal system of point scatterers, In: Novak M (ed) Paradigms of complexity: fractals and structures in the sciences. World Science, Singapore, pp 41–49
18. Uchaikin VV (2004) If the universe were a Levy-Mandelbrot fractal..., gravitation and cosmology 10(1–2):38–38;5–24
19. Uchakin VV (2004) The mesofractal universe driven by Rayleigh-Levy walks. Gen Relativ Gravit 36(7):1689–1717
20. Uchaikin VV (2013) Fractional phenomenology of cosmic ray anomalous diffusion. Physics - Uspekhi 56(11):1074
21. Uchaikin VV, Chernjaev GV (1989) Phenomenological analysis of spatial distributions of shower particles. Yadernaja Fizika 50(9):736–742 (in Russian)
22. Uchaikin VV, Ryzhov VV (1988) Stochastic theory of high-energy particle transport. Nauka, Sibirian Branch, Novosibirsk (in Russian)
23. Sevastyanov BA (1971) Branching processes, M., Nauka, (in Russian)
24. von Holtsmark J (1919) Uber die Verbreiterung von Spektrallinien. Annalen der Physik 58:577–630
25. Chandrasekhar S, von Neumann J (1941) The statistics of the gravitational field arising from a random distribution of stars I. Astrophys J 95:489–531
26. Chandrasekhar S, von Neumann J (1943) The statistics of the gravitational field arising from a random distribution of stars II. Astrophys J 97:1–27
27. Chandrasekhar S (1944) The statistics of the gravitational field arising from a random distribution of stars III. Astrophys J 99:25–46
28. Chandrasekhar S (1944) The statistics of the gravitational field arising from a random distribution of stars IV. Astrophys J 99:47–58
29. Uchaikin VV, Zolotarev VM (1999) Chance and stability. Stable distribution and their applications. Utrecht, the Netherlands, VSP
30. Mikhailov GA (1987) Optimization of Monte Carlo weighting methods. Nauka, Moscow (in Russian)
31. Levermore CD, Pomraning GC, Sanzo DL (1986) Linear transport theory in a random medium. J Math Phys 27:2526–2536

32. Pomraning GC (1989) The Milne problem in a statistical medium. J Quant Spectrosc Radiat Transfer 41:103–115
33. Vanderhaegen D (1986) Radiative transfer in statistically heterogeneous mixtures. Ibid 36:551–561
34. Vanderhaegen D (1988) Impact of a mixing structure on radiative transfer in random media. Ibid 39:333–337
35. Sahni DC (1989) An application of reactor noise techniques to neutron transport problem in a random medium. Ann Nucl Energy 16:397–408
36. Sahni DC (1989) Equvalence of generic equation method and phenomenological model for linear transport problem in a two-state random scattering medium. J Math Phys 30:1554–1559
37. Levermore CD, Wong J, Pomraning O (1988) Renewal theory for transport processes in binary statistical mixtures. J Math Phys 29(4):995–1004
38. Sanchez R (1989) Linear kinetic theory in stochastic media. J Math Phys 30:2498–2511

Kinetic Equation for Systems with Resonant Captures and Scatterings

A. V. Artemyev, A. I. Neishtadt, and A. A. Vasiliev

Abstract We study a Hamiltonian system of type describing a charged particle resonant interaction with an electromagnetic wave. We consider an ensemble of particles that repeatedly pass through the resonance with the wave, and study evolution of the distribution function due to multiple scatterings on the resonance and trappings (captures) into the resonance. We derive the corresponding kinetic equation. Particular cases of this problem has been studied in our recent papers [1, 2].

1 Introduction

Resonant phenomena are a key part in long-term evolution of numerous systems in plasma physics, hydrodynamics, celestial mechanics, etc. The phenomena of scattering on a resonance and capture (trapping) into a resonance were described in details in [3, 4] (see also [5, 6]), and all the characteristics of a single passage through a resonance were obtained. These results were applied to studies of the resonant phenomena in various problems in physics; among recent studies we just mention papers [7–12]. However, in physical systems one has usually to deal with an ensemble of particles (phase trajectories), which pass repeatedly through the resonance during long time intervals. These multiple resonant interactions affect the distribution function of the ensemble. Thus a crucial issue is to implement the properties of individual resonant interactions into a kinetic description of evolution of the distribution function.

A. V. Artemyev
Institute of Geophysics and Planetary Physics, UCLA, Los Angeles, CA, USA
e-mail: aartemyev@igpp.ucla.edu

A. I. Neishtadt
Department of Mathematical Sciences, Loughborough University,
Loughborough LE11 3TU, UK

A. V. Artemyev · A. I. Neishtadt · A. A. Vasiliev (✉)
Space Research Institute, RAS, Moscow, Russia
e-mail: avasiliev@gmail.com; valex@iki.rssi.ru

© Higher Education Press 2021
D. Volchenkov (ed.), *Nonlinear Dynamics, Chaos, and Complexity*,
Nonlinear Physical Science, https://doi.org/10.1007/978-981-15-9034-4_10

A major peculiarity on this way is that captures into resonances provide fast and large-distance transport in the phase space, which cannot be described with differential operators in the kinetic equation. In the papers [13–15], it was proposed to introduce integral operators describing this kind of transport. This approach, however, did not take into account kinetic balance between the captures and the scattering. Namely, while rare captures result in strong variation (say, growth) of energy of a small part of particles (phase trajectories), scatterings produce small energy variation in the opposite direction (decrease) of a large sub-ensemble. Therefore, to include these phenomena into the kinetic equation, one should find and implement the relationship between the corresponding kinetic coefficients. This approach was first proposed in [1] in the simplest case of a Hamiltonian system with one and a half d.o.f., and in [2] for a more realistic system with two d.o.f. In these papers, we have introduced a Fokker-Planck kinetic equation describing evolution of an ensemble of particles in a system where repeated scatterings on resonances and captures into resonances (followed by escapes from the resonances) take place. Our approach is based on the fact that one can introduce probability of capture into a resonance, and that this probability turns out to be interconnected with the velocity of the drift in the phase space due to scatterings on the resonance.

In the present work we derive the kinetic equation in a general case when the time period between successive passages through the resonance depends on the particle energy. In Sect. 2, we briefly outline the main approaches and results concerning an individual resonance crossing. In Sect. 3, we use these results to construct the kinetic equation describing the long-term evolution of the distribution function in a system with multiple resonant captures and scatterings. Note that in [2] the similar equation was obtained with smaller terms omitted. In the present paper, these terms are taken into account allowing to represent the kinetic equation in a more elegant form.

2 Resonant Phenomena in Slow-Fast Hamiltonian Systems

Consider a Hamiltonian system with Hamiltonian

$$H = H_0(p, q) + \varepsilon A(p, q) \sin(kq - \omega t), \tag{1}$$

where ε is a small parameter and (p, q) are canonically conjugate variables. Such Hamiltonians naturally appear in problems of motion of a charged particle in a harmonic electromagnetic wave and a background magnetic field. This is a Hamiltonian system with $1\frac{1}{2}$ degrees of freedom. Introduce t as a new canonical coordinate u, and U as the canonically conjugate momentum. The Hamiltonian takes the form

$$H = U + H_0(p, q) + \varepsilon A(p, q) \sin(kq - \omega u).$$

Now we introduce the phase of the wave as an independent variable $\varphi = q - \omega u/k$. To do this, we make a canonical transformation $(p, q, U, u) \mapsto (\hat{p}, \hat{q}, I, \varphi)$ using generating function

$$W = I\left(q - \frac{\omega}{k}u\right) + \hat{p}q + \hat{U}u$$

Omitting constant $\hat{U}$ and omitting hats over p and q, we obtain a 2 degrees of freedom Hamiltonian (we keep the same notations for the functions H_0 and A):

$$H = -\frac{\omega}{k}I + H_0(p, q, I) + \varepsilon A(p, q, I)\sin(k\varphi) \tag{2}$$

Now we rescale the variables introducing $\bar{\varphi} = k\varphi$. In order to keep the symplectic structure, we also rescale time introducing $\bar{t} = kt$ and consider (p, kq) as a pair of canonically conjugate variables. We assume that $k^{-1} = \varepsilon$. Omitting the bars we obtain the Hamiltonian in the new variables:

$$H = H_0(p, q, I) - v_\phi I + \varepsilon A(p, q, I)\sin \varphi, \tag{3}$$

where we have used the notation $v_\phi = \omega/k$. One can see from (3) that with the accuracy of order $\sim \varepsilon$ the system stays on the energy level $H_0(p, q, I) - v_\phi I = \text{const}$; thus, we obtain the following relation between the particle energy $h = H_0$ and the value of I:

$$h - v_\phi I = \text{const.} \tag{4}$$

In Hamiltonian (3), the pairs of conjugate variables are $(p, \varepsilon^{-1}q)$ and (I, φ). The equations of motion in the main approximation are

$$\dot{p} = -\varepsilon\frac{\partial H_0}{\partial q}$$

$$\dot{q} = \varepsilon\frac{\partial H_0}{\partial p} \tag{5}$$

$$\dot{I} = -\varepsilon A \cos \varphi$$

$$\dot{\varphi} = \frac{\partial H_0}{\partial I} - v_\phi.$$

Thus, in this system variable φ is a fast phase, and the other variables are slow. Far from the resonance $\dot{\varphi} = 0$ the equations of motion can be averaged over the fast phase. Thus we obtain the averaged system:

$$\dot{p} = -\varepsilon\frac{\partial H_0}{\partial q}, \quad \dot{q} = \varepsilon\frac{\partial H_0}{\partial p}, \quad \dot{I} = 0 \tag{6}$$

Variable I is the integral of the averaged system (6) and hence is an adiabatic invariant of the exact system (3) (see, e.g., [5]). Far from the resonance, it is preserved with a good accuracy along phase trajectories of (3).

We assume that the slow motion on the (p, q)-plane in the averaged system (6) is periodic. The area bounded by a trajectory of this averaged motion can be considered as a function of the energy $H_0 = h$ or of the corresponding value of I (see 4). The condition of resonance $\partial H_0/\partial I = 0$ defines a curve on the (p, q)-plane (the resonant curve). In a general situation, trajectories of the averaged system cross the resonant curve.

In a small vicinity of the resonance, the averaging of Eq. (5) does not work properly, and here we apply the standard approach developed in [3] (see also, e.g., [5, 6]). We expand the Hamiltonian H into series near the resonant value of $I = I_R$, where $I_R = I_R(p, q)$ is found from the equation $\partial H_0/\partial I = v_\phi$. Thus we obtain Hamiltonian

$$H = \Lambda(p, q) + \frac{1}{2}g(p, q)(I - I_R)^2 + \varepsilon A(p, q, I_R) \sin \varphi, \qquad (7)$$

where $\Lambda = H_0(p, q, I_R)$ and $g = \partial^2 H_0/\partial I^2\big|_{I=I_R}$ and smaller terms are omitted. Introduce new canonical momentum $K = I - I_R$ with the generating function $W = \bar{p}\varepsilon^{-1}q + (K + I_R)\varphi$, where $(\bar{p}, \bar{q})$ are new variables. In the new variables the Hamiltonian takes the form (bars are omitted, we keep the same notations for the functions Λ and A):

$$H = \Lambda(p, q) + \frac{1}{2}g(p, q)K^2 + \varepsilon A(p, q) \sin \varphi + \varepsilon\beta(p, q)\varphi \equiv \Lambda(p, q) + F, \quad (8)$$

where $\beta(p, q) = \{I_R, \Lambda\}$, $\{\cdot, \cdot\}$ denotes the Poisson bracket with respect to (p, q), and we have introduced the so-called pendulum-like Hamiltonian F. The coefficients g, A, and β in F depend on slow variables p, q, while the evolution of p, q is defined by Hamiltonian Λ. If $A(p, q) > \beta(p, q)$, the phase portrait of F on the (φ, K)-plane has a saddle point and a separatrix, see Fig. 1. The area S of the region inside the separatrix loop can be found as

$$S(p, q) = 2\int_{\varphi_{\min}}^{\varphi_1} K\,d\varphi = \sqrt{\varepsilon}\int_{\varphi_{\min}}^{\varphi_1} \sqrt{\frac{2}{g}\left(\frac{F_s}{\varepsilon} - A\sin\varphi - \beta\varphi\right)}\,d\varphi, \qquad (9)$$

where F_s is the value of F at the saddle point, φ_1 and $\varphi_{\min}$ are shown in Fig. 1.

Closed phase trajectories on the phase portrait of the pendulum-like Hamiltonian F correspond to phase points captured into the resonance, while open trajectories correspond to those passing through the resonance. If S grows, there appears additional phase volume inside of the separatrix loop, and phase points can be captured into the resonance. Motion on the phase portrait is fast compared to the speed of variation of p, q. Hence, the area surrounded by a captured trajectory is an adiabatic invariant of this system. Therefore, while the area S grows, the phase point stays within the sep-

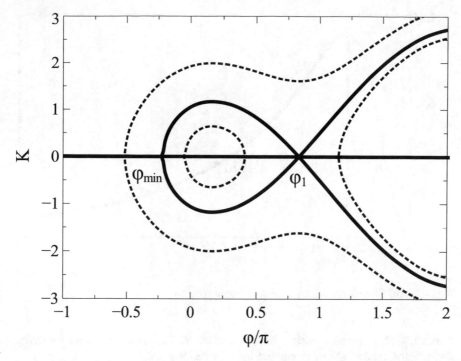

Fig. 1 Phase portrait of Hamiltonian F in (8) in the case $A(p, q) > \beta(p, q)$. It is assumed that g, β are positive

aratrix loop. If later S decreases, the phase point can leave the separatrix loop when the area S again equals the same value as at the time of capture. This is an escape from the resonance. Hence to predict the escape from the resonance one can use the time profile of the function $S(p, q) = S(t)$ along the resonant trajectory where the evolution of (p, q) is defined by the Hamiltonian Λ. On the other hand, capture into the resonance is possible only if the phase point approaches the resonance when the function $S(t)$ grows.

While a phase point is captured, the corresponding value h of the Hamiltonian H_0 of the averaged system (6) varies with time. The value h can be used to parametrize function S, and it is useful to consider S as a function of h: $S = S(h)$. We assume that $S(h)$ has the only maximum at $h = h_{max}$ (see Fig. 2). Thus, phase points captured at $h_- < h_{max}$ are transported in Fig. 2 to the right and escape from the resonance at $h_+ < h_{max}$ such that $S(h_+) = S(h_-)$. One can see that a capture followed by escape from the resonance result in strong (of order 1) variation of the value of h (and of the value of I, see 4).

Capture into a resonance is a probabilistic process (see [3]). Consider a small time interval Δt. The probability of capture can be calculated as the ratio of the number of phase points captured into the resonance during this interval (i.e., $\sim \Delta t \dot{S}$) to the total

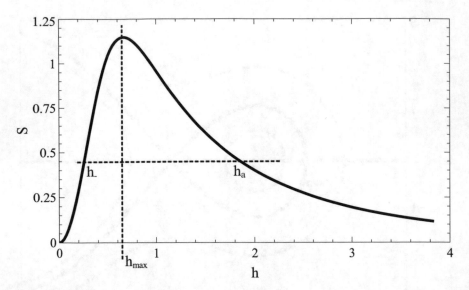

Fig. 2 Plot of the area S as a function of the particle energy h

number of phase points crossing the resonant curve. Thus one obtains the following formula for the probability of capture into the resonance:

$$\Pi = \frac{\{S, \Lambda\}}{2\pi |\beta|}, \quad \text{if } \{S, \Lambda\} > 0,$$

$$\Pi = 0, \quad \text{if } \{S, \Lambda\} \leq 0.$$

(10)

One can see from (10) and (9) that the capture probability is a small value of order $\sqrt{\varepsilon}$.

Phase points that cross the resonant curve without capture are scattered on the resonance. The scattering results in a small variation $\Delta I \sim \sqrt{\varepsilon}$. Exact amplitude of scattering is a random value (see, e.g., [3, 6]). If we have an ensemble of phase points, the mean scattering amplitude is (see [3])

$$\langle \Delta I \rangle = -\mathrm{sign}(\beta) \frac{S}{2\pi},$$

(11)

where $\langle \cdot \rangle$ denotes the ensemble average. In terms of the particle energy h, the mean scattering amplitude is

$$\langle \Delta h \rangle = -\mathrm{sign}(\beta) \frac{S}{2\pi} v_\phi.$$

(12)

To summarize, suppose we have an ensemble of phase points with the same initial value of h. After crossing the resonance, a small part of this ensemble given by (10) is captured into the resonance and its energy significantly changes. The other phase points of the original ensemble are scattered on the resonance with the mean variation of energy given by (12). Generally speaking, on each period $\tau(h)$ of the slow motion a phase trajectory of the averaged system crosses the resonance several times. Assume for simplicity that $A \neq 0$ at only one of these crossings. (Such situations occur in physical problems, see, e.g., [2]). Repeated passages through the resonance result in drift and diffusion of h. Introduce the drift velocity and the diffusion coefficient as

$$V_h = \langle \Delta h \rangle / \tau(h), \quad D_{hh} = \langle (\Delta h)^2 \rangle / \tau(h). \tag{13}$$

Next step is to establish the relation between the capture probability Π and the drift velocity V_h. From (10) and (4) one obtains (if $\{S, \Lambda\} > 0$):

$$\Pi = \frac{1}{2\pi |\beta|} \frac{dS}{dh} \frac{dh}{dI} \frac{dI}{dt} \bigg|_{I=I_R} = \frac{1}{2\pi |\beta|} \frac{dS}{dh} v_\phi \beta = \frac{v_\phi}{2\pi} \text{sign}(\beta) \frac{dS}{dh}. \tag{14}$$

Comparing this expression with (12), we find

$$\Pi = -\frac{d\langle \Delta h \rangle}{dh}. \tag{15}$$

3 Evolution of the Distribution Function

Consider the distribution function of the phase points $f(h, t)$. The kinetic equation for this distribution function has a general form

$$\frac{\partial f}{\partial t} = L_s f + L_c f, \tag{16}$$

where operators L_s and L_c are related to scattering and capture/escape processes, respectively. The scattering part has a standard form

$$L_s f = -\frac{\partial (f V_h)}{\partial h} + \frac{1}{2} \frac{\partial}{\partial h} \left(D_{hh} \frac{\partial f}{\partial h} \right) + L_{sm} f. \tag{17}$$

Here V_h, D_{hh} are drift and diffusion coefficients respectively, defined in the previous section, and L_{sm} is an additional small ($\sim D_{hh}$) drift term. This term appears because V_h is calculated in the principal order in $\sqrt{\varepsilon}$, and it will be omitted in the following consideration.

We assume that the function $S(h)$ has only one maximum at $h = h_{\max}$. The capture/escape operator in (16) has different forms for $h < h_{\max}$ (capture) and $h > h_{\max}$ (escape from the resonance). In the case of capture, $h < h_{\max}$, we have

$$L_c f = -\frac{\Pi(h) f}{\tau}, \tag{18}$$

where $\Pi(h)$ is the probability of capture and $\tau = \tau(h)$ is the period of the averaged motion. Using (15) we find from (18)

$$L_c f = \frac{f}{\tau} \frac{d\langle \Delta h \rangle}{dh}. \tag{19}$$

In the case of escape, $h > h_{\max}$, introduce h_* as the value of the energy that the phase point had before the capture to escape with energy h. Denote $\Pi_* = \Pi(h_*)$, $\tau_* = \tau(h_*)$, $f_* = f(h_*, t)$. Then we have

$$
\begin{aligned}
L_c f &= \frac{\Pi_* f_*}{\tau_*} \left| \frac{dh_*}{dh} \right| = -\frac{\Pi_* f_*}{\tau_*} \frac{dh_*}{dh} = -\frac{\Pi_*}{\tau_*} \frac{dS(h)/dh}{dS(h_*)/dh_*} f_* \\
&= -\frac{v_\phi}{\tau_*} \mathrm{sign}(\beta) \frac{dS(h_*)/dh_*}{2\pi} \frac{dS(h)/dh}{dS(h_*)/dh_*} f_* \\
&= -\frac{v_\phi}{\tau_*} \mathrm{sign}(\beta) \frac{dS(h)/dh}{2\pi} f_* = \frac{d\langle \Delta h \rangle}{dh} \frac{f_*}{\tau_*}.
\end{aligned}
$$

Substituting the above expressions into (16) and using (13) we obtain the following form of the kinetic equation:

At $h < h_{\max}$

$$\frac{\partial f}{\partial t} = -V_h \frac{\partial f}{\partial h} + \frac{1}{\tau} \frac{\partial \tau}{\partial h} V_h f + \frac{1}{2} \frac{\partial}{\partial h} \left(D_{hh} \frac{\partial f}{\partial h} \right); \tag{20}$$

at $h > h_{\max}$

$$\frac{\partial f}{\partial t} = -V_h \frac{\partial f}{\partial h} - \frac{\partial V_h}{\partial h} \left(f - f_* \frac{\tau}{\tau_*} \right) + \frac{1}{\tau_*} \frac{\partial \tau}{\partial h} V_h f_* + \frac{1}{2} \frac{\partial}{\partial h} \left(D_{hh} \frac{\partial f}{\partial h} \right). \tag{21}$$

In [2], we omitted smaller terms with $\tau^{-1} \partial \tau / \partial h$ in Eqs. (20), (21). This does not affect significantly the numerical results. However, now we keep these terms to proceed to a more concise form of the kinetic equation.

One can rewrite kinetic Eqs. (20), (21) using the action variable of the averaged system J instead of the energy h. According to the Hamiltonian equations of motion, these two variables are interconnected via $\partial h / \partial J = 2\pi / \tau$. Using this relation we introduce $\tilde{f}(J, t)$, V_J, D_{JJ} in place of $f(h, t)$, V_h, D_{hh} in the kinetic equation and take into account that

$$f = \frac{\tilde{f} \tau}{2\pi}, \quad V_h = \frac{2\pi V_J}{\tau}, \quad D_{hh} = \frac{4\pi^2 D_{JJ}}{\tau^2}. \tag{22}$$

After straightforward calculations we finally obtain the kinetic equation in terms of the action J (we omitted tildes over f):

At $h < h_{\max}$

$$\frac{\partial f}{\partial t} = -V_J \frac{\partial f}{\partial J} + \frac{1}{2} \frac{\partial}{\partial J} \left(D_{JJ} \frac{\partial f}{\partial J} \right);$$ (23)

at $h > h_{\max}$

$$\frac{\partial f}{\partial t} = -V_J \frac{\partial f}{\partial J} - \frac{\partial V_J}{\partial J} (f - f_*) + \frac{1}{2} \frac{\partial}{\partial J} \left(D_{JJ} \frac{\partial f}{\partial J} \right).$$ (24)

One can find numerical evidence supporting validity of kinetic Eqs. (20–24) in our paper [2].

Acknowledgements The work was supported by the Russian Scientific Foundation, Project No. 19-12-00313.

References

1. Artemyev AV, Neishtadt AI, Vasiliev AA, Mourenas D (2016) Kinetic equation for nonlinear resonant wave-particle interaction. Phys Plasmas 23:090701
2. Artemyev AV, Neishtadt AI, Vasiliev AA, Mourenas D (2017) Probabilistic approach to nonlinear wave-particle resonant interaction. Phys Rev E 95:023204
3. Neishtadt AI (1999) Hamiltonian systems with three or more degrees of freedom, NATO ASI Series C. 533:193–213; Kluwer Academic Publishers, Dordrecht. https://doi.org/10.1063/1.166236
4. Neishtadt A (1975) Passage through a separatrix in a resonance problem with a slowly-varying parameter. J Appl Math Mech 39:594–605. https://doi.org/10.1016/0021-8928(75)90060-X
5. Arnold VI, Kozlov VV, Neishtadt AI (2006) Mathematical aspects of classical and celestial mechanics, 3rd edn. Dynamical systems III. Encyclopedia of mathematical sciences. Springer, New York
6. Neishtadt AI, Vasiliev AA (2006) Destruction of adiabatic invariance at resonances in slow-fast Hamiltonian systems. Nucl Instr Meth Phys Res A 561:158
7. Itin AP, Neishtadt AI, Vasiliev AA (2000) Captures into resonance and scattering on resonance in dynamics of a charged relativistic particle in magnetic field and electrostatic wave. Physica D: Nonlinear Phenomena 141:281–296. https://doi.org/10.1016/S0167-2789(00)00039-7
8. Vasiliev A, Neishtadt A, Artemyev A (2011) Nonlinear dynamics of charged particles in an oblique electromagnetic wave. Phys Lett A 375:3075–3079. https://doi.org/10.1016/j.physleta.2011.06.055
9. Neishtadt A, Vasiliev A, Artemyev A (2011) Resonance-induced surfatron acceleration of a relativistic particle. Moscow Math J 11(3):531–545
10. Vainchtein D, Mezić I (2004) Capture into resonance: a method for efficient control. Phys Rev Lett 93(8):084301. https://doi.org/10.1103/PhysRevLett.93.084301
11. Artemyev AV, Neishtadt AI, Zelenyi LM, Vainchtein DL (2010) Adiabatic description of capture into resonance and surfatron acceleration of charged particles by electromagnetic waves. Chaos 20(4):043128. https://doi.org/10.1063/1.3518360
12. Artemyev AV, Vasiliev AA (2015) Resonant ion acceleration by plasma jets: effects of jet breaking and the magnetic-field curvature. Phys Rev E 91(5):053104. https://doi.org/10.1103/PhysRevE.91.053104
13. Shklyar DR (1981) Stochastic motion of relativistic particles in the field of a monochromatic wave. Sov Phys JETP 53:1187–1192

14. Artemyev AV, Vasiliev AA, Mourenas D, Agapitov O, Krasnoselskikh V, Boscher D, Rolland G (2014) Fast transport of resonant electrons in phase space due to nonlinear trapping by whistler waves. Geophys Res Lett 41:5727–5733. https://doi.org/10.1002/2014GL061380
15. Omura Y, Miyashita Y, Yoshikawa M, Summers D, Hikishima M, Ebihara Y, Kubota Y (2015) Formation process of relativistic electron flux through interaction with chorus emissions in the Earth's inner magnetosphere. J Geophys Res 120:9545–9562. https://doi.org/10.1002/2015JA021563

Solvability in the Sense of Sequences for Some Non-Fredholm Operators in Higher Dimensions

Vitali Vougalter and Vitaly Volpert

Abstract We study solvability of certain linear nonhomogeneous elliptic problems and establish that under reasonable technical assumptions the convergence in $L^2(\mathbb{R}^d)$ of their right sides implies the existence and the convergence in $H^1(\mathbb{R}^d)$ of the solutions. The equations involve the square roots of the sums of second order non-Fredholm differential operators and we rely on the methods of the spectral and scattering theory for Schrödinger type operators similarly to our earlier work [26].

Keywords Solvability conditions · Non-fredholm operators · Sobolev spaces

AMS Subject Classification: 35J10 · 35P10 · 35P25 · 47F05

1 Introduction

Consider the problem

$$\sqrt{-\Delta + V(x)}u - au = f, \tag{1}$$

where $u \in E = H^1(\mathbb{R}^d)$ and $f \in F = L^2(\mathbb{R}^d)$, $d \in \mathbb{N}$, a is a constant and the scalar potential function $V(x)$ tends to 0 at infinity. For $a \geq 0$, the essential spectrum of the operator $A : E \to F$ which corresponds to the left side of problem (1) contains the origin. Consequently, this operator fails to satisfy the Fredholm property. Its image is not closed, for $d > 1$ the dimensions of its kernel and the codimension of its image are not finite. The present work deals with the studies of certain properties of the operators of this kind. Let us recall that elliptic equations containing non Fredholm operators were treated extensively in recent years (see [15–17, 19–25], also [6])

V. Vougalter (✉)
Department of Mathematics, University of Toronto, Toronto, ON M5S 2E4, Canada
e-mail: vitali@math.toronto.edu

V. Volpert
Institute Camille Jordan, UMR 5208 CNRS, University Lyon 1, Villeurbanne 69622, France
e-mail: volpert@math.univ-lyon1.fr

© Higher Education Press 2021
D. Volchenkov (ed.), *Nonlinear Dynamics, Chaos, and Complexity*,
Nonlinear Physical Science, https://doi.org/10.1007/978-981-15-9034-4_11

along with their potential applications to the theory of reaction-diffusion problems (see [8, 9]). Non-Fredholm operators are also important when studying wave systems with an infinite number of localized traveling waves (see [1]). In the particular case when $a = 0$ the operator A^2 satisfies the Fredholm property in some properly chosen weighted spaces (see [2–6]). However, the case of $a \neq 0$ is significantly different and the approach developed in these works cannot be applied.

One of the important issues about problems with non-Fredholm operators concerns their solvability. We address it in the following setting. Let f_n be a sequence of functions in the image of the operator A, such that $f_n \to f$ in $L^2(\mathbb{R}^d)$ as $n \to \infty$. Denote by u_n a sequence of functions from $H^1(\mathbb{R}^d)$ such that

$$Au_n = f_n, \ n \in \mathbb{N}.$$

Since the operator A does not satisfy the Fredholm property, the sequence u_n may not be convergent. Let us call a sequence u_n such that $Au_n \to f$ a solution in the sense of sequences of problem $Au = f$ (see [15]). If such sequence converges to a function u_0 in the norm of the space E, then u_0 is a solution of this equation. Solution in the sense of sequences is equivalent in this case to the usual solution. However, in the case of the non-Fredholm operators, this convergence may not hold or it can occur in some weaker sense. In such case, solution in the sense of sequences may not imply the existence of the usual solution. In the present work we will find sufficient conditions of equivalence of solutions in the sense of sequences and the usual solutions. In the other words, the conditions on sequences f_n under which the corresponding sequences u_n are strongly convergent. Solvability in the sense of sequences for the sums of Schrödinger type operators without Fredholm property was treated in [26].

In the first part of the article we study the problem

$$\sqrt{-\Delta_x + V(x) - \Delta_y + U(y)}u - au = f(x, y), \quad x, y \in \mathbb{R}^3. \tag{2}$$

The operator

$$H_{U, V} := \sqrt{-\Delta_x + V(x) - \Delta_y + U(y)} \tag{3}$$

here is defined via the spectral calculus. Here and further down the Laplacian operators Δ_x and Δ_y are with respect to the x and y variables respectively, such that cumulatively $\Delta := \Delta_x + \Delta_y$. Similarly, for the gradients

$$\nabla := \nabla_x + \nabla_y,$$

where ∇_x and ∇_y act on x and y variables respectively. The square roots of second order differential operators are actively used, for instance in the studies of the superdiffusion problems (see e.g., [27] and the references therein), in relativistic Quantum Mechanics (see e.g., [18]). The scalar potential functions involved in (3) are assumed to be shallow and short-range, satisfying the assumptions analogous to the ones of [19] and [21].

Assumption 1 The potential functions $V(x), U(y) : \mathbb{R}^3 \to \mathbb{R}$ satisfy the bounds

$$|V(x)| \leq \frac{C}{1 + |x|^{3.5+\varepsilon}}, \quad |U(y)| \leq \frac{C}{1 + |y|^{3.5+\varepsilon}}$$

with some $\varepsilon > 0$ and $x, y \in \mathbb{R}^3$ a.e. such that

$$4^{\frac{1}{9}} \frac{9}{8} (4\pi)^{-\frac{2}{3}} \|V\|_{L^\infty(\mathbb{R}^3)}^{\frac{1}{9}} \|V\|_{L^{\frac{4}{3}}(\mathbb{R}^3)}^{\frac{8}{9}} < 1, \tag{4}$$

$$4^{\frac{1}{9}} \frac{9}{8} (4\pi)^{-\frac{2}{3}} \|U\|_{L^\infty(\mathbb{R}^3)}^{\frac{1}{9}} \|U\|_{L^{\frac{4}{3}}(\mathbb{R}^3)}^{\frac{8}{9}} < 1 \tag{5}$$

and

$$\sqrt{c_{HLS}} \|V\|_{L^{\frac{3}{2}}(\mathbb{R}^3)} < 4\pi, \quad \sqrt{c_{HLS}} \|U\|_{L^{\frac{3}{2}}(\mathbb{R}^3)} < 4\pi.$$

Here and below C denotes a finite positive constant and c_{HLS} given on p. 98 of [12] is the constant in the Hardy-Littlewood-Sobolev inequality

$$\left| \int_{\mathbb{R}^3} \int_{\mathbb{R}^3} \frac{f_1(x) f_1(y)}{|x - y|^2} dx dy \right| \leq c_{HLS} \|f_1\|_{L^{\frac{3}{2}}(\mathbb{R}^3)}^2, \quad f_1 \in L^{\frac{3}{2}}(\mathbb{R}^3).$$

The norm of a function $f_1 \in L^p(\mathbb{R}^d)$, $1 \leq p \leq \infty$, $d \in \mathbb{N}$ is denoted as $\|f_1\|_{L^p(\mathbb{R}^d)}$. We designate the inner product of two functions as

$$(f(x), g(x))_{L^2(\mathbb{R}^d)} := \int_{\mathbb{R}^d} f(x) \bar{g}(x) dx, \tag{6}$$

with a slight abuse of notations when such functions are not square integrable. Indeed, if $f(x) \in L^1(\mathbb{R}^d)$ and $g(x)$ is bounded like, for example the functions of the continuos spectrum of the Schrödinger operators discussed below, then the integral in the right side of (6) is well defined. By means of Lemma 2.3 of [21], under Assumption 1 above on the scalar potentials, operator (3) considered as acting in $L^2(\mathbb{R}^6)$ with domain $H^1(\mathbb{R}^6)$ is self-adjoint and is unitarily equivalent to $\sqrt{-\Delta_x - \Delta_y}$ on $L^2(\mathbb{R}^6)$ via the product of the wave operators (see [11, 14])

$$\Omega_V^{\pm} := s - \lim_{t \to \mp\infty} e^{it(-\Delta_x + V(x))} e^{it\Delta_x}, \quad \Omega_U^{\pm} := s - \lim_{t \to \mp\infty} e^{it(-\Delta_y + U(y))} e^{it\Delta_y},$$

with the limits here understood in the strong L^2 sense (see e.g., [13] p. 34, [7] p. 90). Hence, operator (3) has no nontrivial $L^2(\mathbb{R}^6)$ eigenfunctions. Its essential spectrum fills the nonnegative semi-axis $[0, +\infty)$. Therefore, operator (3) does not satisfy the Fredholm property. The functions of the continuos spectrum of the first operator involved in (3) are the solutions the Schrödinger equation

$$[-\Delta_x + V(x)]\varphi_k(x) = k^2\varphi_k(x), \quad k \in \mathbb{R}^3,$$

in the integral form the Lippmann-Schwinger equation

$$\varphi_k(x) = \frac{e^{ikx}}{(2\pi)^{\frac{3}{2}}} - \frac{1}{4\pi} \int_{\mathbb{R}^3} \frac{e^{i|k||x-y|}}{|x-y|}(V\varphi_k)(y)dy \tag{7}$$

and the orthogonality conditions $(\varphi_k(x), \varphi_{k_1}(x))_{L^2(\mathbb{R}^3)} = \delta(k - k_1)$, $k, k_1 \in \mathbb{R}^3$. The integral operator involved in (7)

$$(Q\varphi)(x) := -\frac{1}{4\pi} \int_{\mathbb{R}^3} \frac{e^{i|k||x-y|}}{|x-y|}(V\varphi)(y)dy, \quad \varphi(x) \in L^\infty(\mathbb{R}^3).$$

We consider $Q : L^\infty(\mathbb{R}^3) \to L^\infty(\mathbb{R}^3)$ and its norm $\|Q\|_\infty < 1$ under Assumption 1 via Lemma 2.1 of [21]. In fact, this norm is bounded above by the k-independent quantity $I(V)$, which is the left side of inequality (4). Analogously, for the second operator involved in (3) the functions of its continuous spectrum solve

$$[-\Delta_y + U(y)]\eta_q(y) = q^2\eta_q(y), \quad q \in \mathbb{R}^3,$$

in the integral formulation

$$\eta_q(y) = \frac{e^{iqy}}{(2\pi)^{\frac{3}{2}}} - \frac{1}{4\pi} \int_{\mathbb{R}^3} \frac{e^{i|q||y-z|}}{|y-z|}(U\eta_q)(z)dz, \tag{8}$$

such that the the orthogonality relations $(\eta_q(y), \eta_{q_1}(y))_{L^2(\mathbb{R}^3)} = \delta(q - q_1)$, $q, q_1 \in \mathbb{R}^3$ hold. The integral operator involved in (8) is

$$(P\eta)(y) := -\frac{1}{4\pi} \int_{\mathbb{R}^3} \frac{e^{i|q||y-z|}}{|y-z|}(U\eta)(z)dz, \quad \eta(y) \in L^\infty(\mathbb{R}^3).$$

For $P : L^\infty(\mathbb{R}^3) \to L^\infty(\mathbb{R}^3)$ its norm $\|P\|_\infty < 1$ under Assumption 1 by virtue of Lemma 2.1 of [21]. As before, this norm can be estimated above by the q-independent quantity $I(U)$, which is the left side of inequality (5). Let us denote by the double tilde sign the generalized Fourier transform with the product of these functions of the continuous spectrum

$$\tilde{\tilde{f}}(k, q) := (f(x, y), \varphi_k(x)\eta_q(y))_{L^2(\mathbb{R}^6)}, \quad k, q \in \mathbb{R}^3. \tag{9}$$

Equation (9) is a unitary transform on $L^2(\mathbb{R}^6)$. We will be using the Sobolev space

$$H^1(\mathbb{R}^d) = \{u(x) : \mathbb{R}^d \to \mathbb{C} \mid u(x) \in L^2(\mathbb{R}^d), \ \nabla u \in L^2(\mathbb{R}^d)\}$$

equipped with the norm

$$\|u\|_{H^1(\mathbb{R}^d)}^2 = \|u\|_{L^2(\mathbb{R}^d)}^2 + \|\nabla u\|_{L^2(\mathbb{R}^d)}^2, \quad d \in \mathbb{N}.$$

Our first main proposition is as follows.

Theorem 2 *Let Assumption 1 hold and $f(x, y) \in L^2(\mathbb{R}^6)$.*

(a) When $a = 0$, let in addition $f(x, y) \in L^1(\mathbb{R}^6)$. Then Eq. (2) admits a unique solution $u(x, y) \in H^1(\mathbb{R}^6)$.

(b) When $a > 0$, let in addition $xf(x, y)$, $yf(x, y) \in L^1(\mathbb{R}^6)$. Then problem (2) possesses a unique solution $u(x, y) \in H^1(\mathbb{R}^6)$ if and only if

$$(f(x, y), \varphi_k(x)\eta_q(y))_{L^2(\mathbb{R}^6)} = 0, \quad (k, q) \in S_a^6. \tag{10}$$

Here and below S_a^d stands for the sphere in $\mathbb{R}^d$ of radius a centered at the origin. Such unit sphere will be denoted as S^d and its Lebesgue measure as $|S^d|$. Note that in the case of $a = 0$ in the theorem above no orthogonality conditions are needed to solve Eq. (2) in $H^1(\mathbb{R}^6)$.

Then we turn our attention to the issue of the solvability in the sense of sequences for our problem. The corresponding sequence of equations with $n \in \mathbb{N}$ is given by

$$\sqrt{-\Delta_x + V(x) - \Delta_y + U(y)}u_n - au_n = f_n(x, y), \quad x, y \in \mathbb{R}^3 \tag{11}$$

with the right sides convergent to the right side of (2) in $L^2(\mathbb{R}^6)$ as $n \to \infty$.

Theorem 3 *Let Assumption 1 hold, $n \in \mathbb{N}$ and $f_n(x, y) \in L^2(\mathbb{R}^6)$, such that $f_n(x, y) \to f(x, y)$ in $L^2(\mathbb{R}^6)$ as $n \to \infty$.*

(a) When $a = 0$, let in addition $f_n(x, y) \in L^1(\mathbb{R}^6)$, $n \in \mathbb{N}$, such that $f_n(x, y) \to f(x, y)$ in $L^1(\mathbb{R}^6)$ as $n \to \infty$. Then equations (2) and (11) have unique solutions $u(x, y) \in H^1(\mathbb{R}^6)$ and $u_n(x, y) \in H^1(\mathbb{R}^6)$ respectively, such that $u_n(x, y) \to u(x, y)$ in $H^1(\mathbb{R}^6)$ as $n \to \infty$.

(b) When $a > 0$, let in addition $xf_n(x, y)$, $yf_n(x, y) \in L^1(\mathbb{R}^6)$, $n \in \mathbb{N}$, such that $xf_n(x, y) \to xf(x, y)$, $yf_n(x, y) \to yf(x, y)$ in $L^1(\mathbb{R}^6)$ as $n \to \infty$ and the orthogonality conditions

$$(f_n(x, y), \varphi_k(x)\eta_q(y))_{L^2(\mathbb{R}^6)} = 0, \quad (k, q) \in S_a^6. \tag{12}$$

hold for all $n \in \mathbb{N}$. Then problems (2) and (11) admit unique solutions $u(x, y) \in H^1(\mathbb{R}^6)$ and $u_n(x, y) \in H^1(\mathbb{R}^6)$ respectively, such that $u_n(x, y) \to u(x, y)$ in $H^1(\mathbb{R}^6)$ as $n \to \infty$.

In the second part of the article we consider the problem

$$\sqrt{-\Delta_x - \Delta_y + U(y)}u - au = \phi(x, y), \quad x \in \mathbb{R}^d, \quad y \in \mathbb{R}^3 \tag{13}$$

with $d \in \mathbb{N}$ and the scalar potential function involved in (13) is shallow and short-range under Assumption 1 as before. The operator

$$L_U := \sqrt{-\Delta_x - \Delta_y + U(y)} \tag{14}$$

here is defined by means of the spectral calculus. Similarly to (3), under our assumptions operator (14) considered as acting in $L^2(\mathbb{R}^{d+3})$ with domain $H^1(\mathbb{R}^{d+3})$ is self-adjoint and is unitarily equivalent to $\sqrt{-\Delta_x - \Delta_y}$. Therefore, operator (14) has no nontrivial $L^2(\mathbb{R}^{d+3})$ eigenfunctions. Its essential spectrum fills the nonnegative semi-axis $[0, +\infty)$ and such that operator (14) fails to satisfy the Fredholm property. Let us consider another generalized Fourier transform with the standard Fourier harmonics and the perturbed plane waves

$$\tilde{\tilde{\phi}}(k, q) := \left(\phi(x, y), \frac{e^{ikx}}{(2\pi)^{\frac{d}{2}}} \eta_q(y) \right)_{L^2(\mathbb{R}^{d+3})}, \quad k \in \mathbb{R}^d, \quad q \in \mathbb{R}^3. \tag{15}$$

Equation (15) is a unitary transform on $L^2(\mathbb{R}^{d+3})$. We have the following statement.

Theorem 4 *Let the potential function $U(y)$ satisfy Assumption 1 and $\phi(x, y) \in L^2(\mathbb{R}^{d+3})$, $d \in \mathbb{N}$.*

(a) When $a = 0$, let in addition $\phi(x, y) \in L^1(\mathbb{R}^{d+3})$. Then Eq. (13) admits a unique solution $u(x, y) \in H^1(\mathbb{R}^{d+3})$.

(b) When $a > 0$, let in addition $x\phi(x, y)$, $y\phi(x, y) \in L^1(\mathbb{R}^{d+3})$. Then problem (13) has a unique solution $u(x, y) \in H^1(\mathbb{R}^{d+3})$ if and only if

$$\left(\phi(x, y), \frac{e^{ikx}}{(2\pi)^{\frac{d}{2}}} \eta_q(y) \right)_{L^2(\mathbb{R}^{d+3})} = 0, \quad (k, q) \in S_a^{d+3}. \tag{16}$$

Note that in the case of $a = 0$ of this theorem no orthogonality relations are needed to solve problem (13) in $H^1(\mathbb{R}^{d+3})$.

Our final main proposition deals with the issue of the solvability in the sense of sequences for our problem. The corresponding sequence of equations with $n \in \mathbb{N}$ is given by

$$\sqrt{-\Delta_x - \Delta_y + U(y)} u_n - a u_n = \phi_n(x, y), \quad x \in \mathbb{R}^d, \quad d \in \mathbb{N}, \quad y \in \mathbb{R}^3 \tag{17}$$

with the right sides convergent to the right side of (13) in $L^2(\mathbb{R}^{d+3})$ as $n \to \infty$.

Theorem 5 *Let the potential function $U(y)$ satisfy Assumption 1, $n \in \mathbb{N}$ and $\phi_n(x, y) \in L^2(\mathbb{R}^{d+3})$, $d \in \mathbb{N}$, such that $\phi_n(x, y) \to \phi(x, y)$ in $L^2(\mathbb{R}^{d+3})$ as $n \to \infty$.*

(a) When $a = 0$, let in addition $\phi_n(x, y) \in L^1(\mathbb{R}^{d+3})$, $n \in \mathbb{N}$, such that $\phi_n(x, y) \to \phi(x, y)$ in $L^1(\mathbb{R}^{d+3})$ as $n \to \infty$. Then equations (13) and (17) possess unique solutions $u(x, y) \in H^1(\mathbb{R}^{d+3})$ and $u_n(x, y) \in H^1(\mathbb{R}^{d+3})$ respectively, such that $u_n(x, y) \to u(x, y)$ in $H^1(\mathbb{R}^{d+3})$ as $n \to \infty$.

(b) When $a > 0$, let in addition $x\phi_n(x, y)$, $y\phi_n(x, y) \in L^1(\mathbb{R}^{d+3})$, such that $x\phi_n(x, y) \to x\phi(x, y)$, $y\phi_n(x, y) \to y\phi(x, y)$ in $L^1(\mathbb{R}^{d+3})$ as $n \to \infty$ and the orthogonality relations

$$\left(\phi_n(x, y), \frac{e^{ikx}}{(2\pi)^{\frac{d}{2}}} \eta_q(y) \right)_{L^2(\mathbb{R}^{d+3})} = 0, \quad (k, q) \in S_a^{d+3}. \tag{18}$$

hold for all $n \in \mathbb{N}$. Then problems (13) and (17) admit unique solutions $u(x, y) \in H^1(\mathbb{R}^{d+3})$ and $u_n(x, y) \in H^1(\mathbb{R}^{d+3})$ respectively, such that $u_n(x, y) \to u(x, y)$ in $H^1(\mathbb{R}^{d+3})$ as $n \to \infty$.

Let us note that (10), (12), (16), (18) are the orthogonality conditions involving the functions of the continuous spectrum of our Schrödinger operators, as distinct from the Limiting Absorption Principle in which one orthogonalizes to the standard Fourier harmonics (see e.g., Lemma 2.3 and Proposition 2.4 of [10]). We proceed to the proof of our statements.

2 Solvability in the Sense of Sequences with Two Potentials

Proof of Theorem 2 Let us note that it is sufficient to solve equation (2) in $L^2(\mathbb{R}^6)$, because its square integrable solution will belong to $H^1(\mathbb{R}^6)$ as well. Indeed, using definition (3) it can be trivially verified that $\|H_{U,\,V}u\|_{L^2(\mathbb{R}^6)}^2$ equals to

$$\|\nabla u\|_{L^2(\mathbb{R}^6)}^2 + \int_{\mathbb{R}^6} V(x)|u(x, y)|^2 dx dy + \int_{\mathbb{R}^6} U(y)|u(x, y)|^2 dx dy, \tag{19}$$

where $u(x, y)$ is a square integrable solution of (2), the scalar potentials $V(x)$ and $U(y)$ are bounded by means of Assumption 1 and $f(x, y) \in L^2(\mathbb{R}^6)$ by virtue of the one of our assumptions. Then (19) yields $\nabla u(x, y) \in L^2(\mathbb{R}^6)$, such that $u(x, y) \in H^1(\mathbb{R}^6)$.

To prove the uniqueness of solutions for our problem, we suppose that Eq. (2) has two square integrable solutions $u_1(x, y)$ and $u_2(x, y)$. Then their difference $w(x, y) := u_1(x, y) - u_2(x, y) \in L^2(\mathbb{R}^6)$ satisfies the equation

$$H_{U,\,V} w = aw.$$

Since operator (3) has no nontrivial square integrable eigenfunctions in the whole space as discussed above, we have $w(x, y) = 0$ a.e. in $\mathbb{R}^6$.

First of all, we consider the case of our theorem when $a = 0$. Let us apply the generalized Fourier transform (9) to both sides of problem (2). This yields

$$\tilde{\tilde{u}}(k, q) = \frac{\tilde{\tilde{f}}(k, q)}{\sqrt{k^2 + q^2}} \chi_{\{\sqrt{k^2+q^2}\le 1\}} + \frac{\tilde{\tilde{f}}(k, q)}{\sqrt{k^2 + q^2}} \chi_{\{\sqrt{k^2+q^2}>1\}} \tag{20}$$

with $k, q \in \mathbb{R}^3$. Here and throughout the paper χ_A will denote the characteristic function of a set $A \subseteq \mathbb{R}^d$. Obviously, the second term in the right side of (20) can be estimated from above in the absolute value by $\tilde{\tilde{f}}(k, q) \in L^2(\mathbb{R}^6)$ due to the one of our assumptions. The first term in the right side of (20) can be easily estimated from above in the absolute value by virtue of Corollary 2.2 of [21] by

$$\frac{1}{(2\pi)^3} \frac{1}{1 - I(V)} \frac{1}{1 - I(U)} \|f\|_{L^1(\mathbb{R}^6)} \frac{\chi_{\{\sqrt{k^2+q^2} \le 1\}}}{\sqrt{k^2 + q^2}}.$$

Therefore,

$$\left\| \frac{\tilde{\tilde{f}}(k, q)}{\sqrt{k^2 + q^2}} \chi_{\{\sqrt{k^2+q^2} \le 1\}} \right\|_{L^2(\mathbb{R}^6)} \le \frac{1}{(2\pi)^3} \frac{1}{1 - I(V)} \frac{1}{1 - I(U)} \|f\|_{L^1(\mathbb{R}^6)} \frac{\sqrt{|S^6|}}{2},$$

which is finite as assumed in the theorem. Hence the unique solution $u(x, y) \in L^2(\mathbb{R}^6)$.

We conclude the proof with treating the case b) of the theorem. We apply the generalized Fourier transform (9) to both sides of Eq. (2) and arrive at

$$\tilde{\tilde{u}}(k, q) = \frac{\tilde{\tilde{f}}(k, q)}{\sqrt{k^2 + q^2} - a}.$$

Let us introduce the set

$$A_\delta := \{(k, q) \in \mathbb{R}^6 \mid a - \delta \le \sqrt{k^2 + q^2} \le a + \delta\}, \quad 0 < \delta < a, \quad (21)$$

such that

$$\tilde{\tilde{u}}(k, q) = \frac{\tilde{\tilde{f}}(k, q)}{\sqrt{k^2 + q^2} - a} \chi_{A_\delta} + \frac{\tilde{\tilde{f}}(k, q)}{\sqrt{k^2 + q^2} - a} \chi_{A_\delta^c}. \quad (22)$$

Note that for a set $A \subseteq \mathbb{R}^d$ we denote its complement as A^c. Evidently, the second term in the right side of (22) can be bounded from above in the absolute value by $\frac{|\tilde{\tilde{f}}(k, q)|}{\delta} \in L^2(\mathbb{R}^6)$ due to the one of our assumptions. Clearly, we have the representation

$$\tilde{\tilde{f}}(k, q) = \tilde{\tilde{f}}(a, \sigma) + \int_a^{\sqrt{k^2+q^2}} \frac{\partial \tilde{\tilde{f}}(s, \sigma)}{\partial s} ds.$$

Here and below σ will denote the angle variables on the sphere. This enables us to express the first term in the right side of (22) as

$$\frac{\tilde{\tilde{f}}(a, \sigma)}{\sqrt{k^2 + q^2} - a} \chi_{A_\delta} + \frac{\int_a^{\sqrt{k^2+q^2}} \frac{\partial \tilde{\tilde{f}}(s,\sigma)}{\partial s} ds}{\sqrt{k^2 + q^2} - a} \chi_{A_\delta}. \quad (23)$$

Evidently, we can estimate the second term in (23) from above in the absolute value by

$$\|(\nabla_k + \nabla_q)\tilde{\tilde{f}}(k, q)\|_{L^\infty(\mathbb{R}^6)}\chi_{A_\delta} \in L^2(\mathbb{R}^6),$$

where the gradients ∇_k and ∇_q act on variables k and q respectively. Note that under our assumptions $(\nabla_k + \nabla_q)\tilde{\tilde{f}}(k, q) \in L^\infty(\mathbb{R}^6)$ by means of Lemma 11 of [19]. Apparently, the first term in (23) is square integrable if and only if $\tilde{\tilde{f}}(a, \sigma)$ vanishes, which is equivalent to orthogonality condition (10). ∎

Let us turn our attention to establishing the solvability in the sense of sequences for our equation in the case of two scalar potentials.

Proof of Theorem 3 Suppose $u(x, y)$ and $u_n(x, y)$, $n \in \mathbb{N}$ are the unique solutions of equations (2) and (11) in $H^1(\mathbb{R}^6)$ with $a \geq 0$ respectively and it is known that $u_n(x, y) \to u(x, y)$ in $L^2(\mathbb{R}^6)$ as $n \to \infty$. Then, it will follow that $u_n(x, y) \to u(x, y)$ in $H^1(\mathbb{R}^6)$ as $n \to \infty$ as well. Indeed, from (2) and (11) we easily derive that

$$H_{U, V}(u_n(x, y) - u(x, y)) = a(u_n(x, y) - u(x, y)) + [f_n(x, y) - f(x, y)],$$

which clearly implies

$$\|H_{U, V}(u_n(x, y) - u(x, y))\|_{L^2(\mathbb{R}^6)} \leq a\|u_n(x, y) - u(x, y)\|_{L^2(\mathbb{R}^6)}+$$

$$+\|f_n(x, y) - f(x, y)\|_{L^2(\mathbb{R}^6)} \to 0, \quad n \to \infty$$

by means of our assumptions. We express

$$\|H_{U, V}(u_n(x, y) - u(x, y))\|^2_{L^2(\mathbb{R}^6)} = \|\nabla(u_n(x, y) - u(x, y))\|^2_{L^2(\mathbb{R}^6)}+$$

$$+\int_{\mathbb{R}^6} V(x)|u_n(x, y) - u(x, y)|^2 dxdy + \int_{\mathbb{R}^6} U(y)|u_n(x, y) - u(x, y)|^2 dxdy$$

with the bounded scalar potentials $V(x)$ and $U(y)$ due to Assumption 1. Thus, in the identity above the left side along with the second and the last term in the right side tend to zero as $n \to \infty$. This yields that $\nabla u_n(x, y) \to \nabla u(x, y)$ in $L^2(\mathbb{R}^6)$ as $n \to \infty$, such that $u_n(x, y) \to u(x, y)$ in $H^1(\mathbb{R}^6)$ as $n \to \infty$ as well.

In the case (a) problems (2) and (11) have unique solutions $u(x, y), u_n(x, y)$ belonging to $H^1(\mathbb{R}^6)$ respectively with $n \in \mathbb{N}$ by virtue of the part (a) of Theorem 2 above. Let us apply the generalized Fourier transform (9) to both sides of equations (2) and (11). This yields

$$\tilde{\tilde{u}}(k, q) = \frac{\tilde{\tilde{f}}(k, q)}{\sqrt{k^2 + q^2}}, \quad \tilde{\tilde{u}}_n(k, q) = \frac{\tilde{\tilde{f}}_n(k, q)}{\sqrt{k^2 + q^2}}, \quad n \in \mathbb{N}.$$

Thus $\tilde{\tilde{u}}_n(k, q) - \tilde{\tilde{u}}(k, q)$ can be written as

$$\frac{\tilde{\tilde{f}}_n(k, q) - \tilde{\tilde{f}}(k, q)}{\sqrt{k^2 + q^2}} \chi_{\{\sqrt{k^2+q^2} \leq 1\}} + \frac{\tilde{\tilde{f}}_n(k, q) - \tilde{\tilde{f}}(k, q)}{\sqrt{k^2 + q^2}} \chi_{\{\sqrt{k^2+q^2} > 1\}}. \qquad (24)$$

Evidently, the second term in (24) can be easily bounded from above in the absolute value by $|\tilde{\tilde{f}}_n(k, q) - \tilde{\tilde{f}}(k, q)|$, such that

$$\left\| \frac{\tilde{\tilde{f}}_n(k, q) - \tilde{\tilde{f}}(k, q)}{\sqrt{k^2 + q^2}} \chi_{\{\sqrt{k^2+q^2} > 1\}} \right\|_{L^2(\mathbb{R}^6)} \leq \| f_n(x, y) - f(x, y) \|_{L^2(\mathbb{R}^6)} \to 0$$

as $n \to \infty$ due to the one of our assumptions. We estimate the first term in (24) from above in the absolute value by means of the Corollary 2.2 of [21] by

$$\frac{1}{(2\pi)^3} \frac{1}{1 - I(V)} \frac{1}{1 - I(U)} \| f_n(x, y) - f(x, y) \|_{L^1(\mathbb{R}^6)} \frac{\chi_{\{\sqrt{k^2+q^2} \leq 1\}}}{\sqrt{k^2 + q^2}},$$

such that

$$\left\| \frac{\tilde{\tilde{f}}_n(k, q) - \tilde{\tilde{f}}(k, q)}{\sqrt{k^2 + q^2}} \chi_{\{\sqrt{k^2+q^2} \leq 1\}} \right\|_{L^2(\mathbb{R}^6)} \leq$$

$$\leq \frac{1}{(2\pi)^3} \frac{1}{1 - I(V)} \frac{1}{1 - I(U)} \| f_n(x, y) - f(x, y) \|_{L^1(\mathbb{R}^6)} \frac{\sqrt{|S^6|}}{2} \to 0, \quad n \to \infty$$

according to the one of our assumptions. Therefore, $u_n(x, y) \to u(x, y)$ in $L^2(\mathbb{R}^6)$ as $n \to \infty$ in the case when the parameter $a = 0$.

Then we proceed to the proof of the part (b) of the theorem. For each $n \in \mathbb{N}$ Eq. (11) admits a unique solution $u_n(x, y) \in H^1(\mathbb{R}^6)$ by means of the result of the part (b) of Theorem 2 above. By virtue of (12) along with Corollary 2.2 of [21], we estimate for $(k, q) \in S_a^6$

$$|(f(x, y), \varphi_k(x)\eta_q(y))_{L^2(\mathbb{R}^6)}| = |(f(x, y) - f_n(x, y), \varphi_k(x)\eta_q(y))_{L^2(\mathbb{R}^6)}| \leq$$

$$\leq \frac{1}{(2\pi)^3} \frac{1}{1 - I(V)} \frac{1}{1 - I(U)} \| f_n(x, y) - f(x, y) \|_{L^1(\mathbb{R}^6)} \to 0, \quad n \to \infty.$$

Note that under our assumptions $f_n(x, y) \to f(x, y)$ in $L^1(\mathbb{R}^6)$ via the simple argument on p. 114 of [26]. Hence, we obtain

$$(f(x, y), \varphi_k(x)\eta_q(y))_{L^2(\mathbb{R}^6)} = 0, \quad (k, q) \in S_a^6. \qquad (25)$$

Therefore, Eq. (2) admits a unique solution $u(x, y) \in H^1(\mathbb{R}^6)$ due to the result of the part (b) of Theorem 2 above. We apply the generalized Fourier transform (9) to both sides of problems (2) and (11). This gives us

$$\tilde{\tilde{u}}_n(k, q) - \tilde{\tilde{u}}(k, q) = \frac{\tilde{\tilde{f}}_n(k, q) - \tilde{\tilde{f}}(k, q)}{\sqrt{k^2 + q^2} - a} \chi_{A_\delta} + \frac{\tilde{\tilde{f}}_n(k, q) - \tilde{\tilde{f}}(k, q)}{\sqrt{k^2 + q^2} - a} \chi_{A_\delta^c} \quad (26)$$

with the set A_δ defined in (21). Clearly, the second term in the right side of (26) can be bounded from above in the absolute value by $\dfrac{|\tilde{\tilde{f}}_n(k, q) - \tilde{\tilde{f}}(k, q)|}{\delta}$, such that

$$\left\| \frac{\tilde{\tilde{f}}_n(k, q) - \tilde{\tilde{f}}(k, q)}{\sqrt{k^2 + q^2} - a} \chi_{A_\delta^c} \right\|_{L^2(\mathbb{R}^6)} \leq \frac{\|f_n(x, y) - f(x, y)\|_{L^2(\mathbb{R}^6)}}{\delta} \to 0, \quad n \to \infty$$

due to the one of our assumptions. Orthogonality conditions (12) and (25) yield

$$\tilde{\tilde{f}}(a, \sigma) = 0, \quad \tilde{\tilde{f}}_n(a, \sigma) = 0, \quad n \in \mathbb{N},$$

such that

$$\tilde{\tilde{f}}(k, q) = \int_a^{\sqrt{k^2 + q^2}} \frac{\partial \tilde{\tilde{f}}(s, \sigma)}{\partial s} ds, \quad \tilde{\tilde{f}}_n(k, q) = \int_a^{\sqrt{k^2 + q^2}} \frac{\partial \tilde{\tilde{f}}_n(s, \sigma)}{\partial s} ds, \quad n \in \mathbb{N}.$$

This enables us to write the first term in the right side of (26) as

$$\frac{\int_a^{\sqrt{k^2 + q^2}} \left[\frac{\partial \tilde{\tilde{f}}_n(s, \sigma)}{\partial s} - \frac{\partial \tilde{\tilde{f}}(s, \sigma)}{\partial s} \right] ds}{\sqrt{k^2 + q^2} - a} \chi_{A_\delta}. \quad (27)$$

Obviously, (27) can be bounded from above in the absolute value by

$$\|(\nabla_k + \nabla_q)(\tilde{\tilde{f}}_n(k, q) - \tilde{\tilde{f}}(k, q))\|_{L^\infty(\mathbb{R}^6)} \chi_{A_\delta}.$$

This allows us to estimate the $L^2(\mathbb{R}^6)$ norm of (27) from above by

$$C\|(\nabla_k + \nabla_q)(\tilde{\tilde{f}}_n(k, q) - \tilde{\tilde{f}}(k, q))\|_{L^\infty(\mathbb{R}^6)} \to 0, \quad n \to \infty$$

by means of the part (a) of Lemma 5 of [26] under our assumptions. Therefore, $u_n(x, y) \to u(x, y)$ in $L^2(\mathbb{R}^6)$ as $n \to \infty$. ∎

In the last section of the article we treat the situation when a free Laplace operator is added to the three dimensional Schrödinger operator.

3 Solvability in the Sense of Sequences with Laplacian and a Single Potential

Proof of Theorem 4 Evidently, it is sufficient to solve problem (13) in $L^2(\mathbb{R}^{d+3})$, since its square integrable solution will belong to $H^1(\mathbb{R}^{d+3})$ as well. Indeed, by means of definition (14) it can be easily verified that $\|L_U u\|^2_{L^2(\mathbb{R}^{d+3})}$ is equal to

$$\|\nabla u\|^2_{L^2(\mathbb{R}^{d+3})} + \int_{\mathbb{R}^{d+3}} U(y)|u(x, y)|^2 dx dy, \quad d \in \mathbb{N} \tag{28}$$

where $u(x, y)$ is a square integrable solution of (13), the scalar potential function $U(y)$ is bounded due to Assumption 1 and $\phi(x, y) \in L^2(\mathbb{R}^{d+3})$ by means of the one of our assumptions. Then (28) implies that $\nabla u(x, y) \in L^2(\mathbb{R}^{d+3})$, such that $u(x, y) \in H^1(\mathbb{R}^{d+3})$.

To establish the uniqueness of solutions for our equation, let us suppose that (13) admits two square integrable solutions $u_1(x, y)$ and $u_2(x, y)$. Then their difference $w(x, y) := u_1(x, y) - u_2(x, y) \in L^2(\mathbb{R}^{d+3})$ is a solution of the equation

$$L_U w = aw.$$

Since operator (14) does not have nontrivial square integrable eigenfunctions in the whole space as mentioned above, we have $w(x, y) = 0$ a.e. in $\mathbb{R}^{d+3}$.

Let us first treat the case of our theorem when $a = 0$. We apply the generalized Fourier transform (15) to both sides of Eq. (13). This yields

$$\tilde{\tilde{u}}(k, q) = \frac{\tilde{\tilde{\phi}}(k, q)}{\sqrt{k^2 + q^2}} \chi_{\left\{\sqrt{k^2+q^2} \leq 1\right\}} + \frac{\tilde{\tilde{\phi}}(k, q)}{\sqrt{k^2 + q^2}} \chi_{\left\{\sqrt{k^2+q^2} > 1\right\}} \tag{29}$$

with $k \in \mathbb{R}^d$, $q \in \mathbb{R}^3$. Clearly, the second term in (29) can be bounded from above in the absolute value by $|\tilde{\tilde{\phi}}(k, q)| \in L^2(\mathbb{R}^{d+3})$ due to the one of our assumptions. Corollary 2.2 of [21] yields

$$|\tilde{\tilde{\phi}}(k, q)| \leq \frac{1}{(2\pi)^{\frac{d+3}{2}}} \frac{1}{1 - I(u)} \|\phi(x, y)\|_{L^1(\mathbb{R}^{d+3})},$$

such that the first term in (29) can be estimated from above in the abosolute value by

$$\frac{1}{(2\pi)^{\frac{d+3}{2}}} \frac{1}{1 - I(u)} \|\phi(x, y)\|_{L^1(\mathbb{R}^{d+3})} \frac{\chi_{\left\{\sqrt{k^2+q^2} \leq 1\right\}}}{\sqrt{k^2 + q^2}}.$$

This implies that

$$\left\| \frac{\tilde{\tilde{\phi}}(k,q)}{\sqrt{k^2+q^2}} \chi_{\{\sqrt{k^2+q^2} \le 1\}} \right\|_{L^2(\mathbb{R}^{d+3})} \le$$

$$\le \frac{1}{(2\pi)^{\frac{d+3}{2}}} \frac{1}{1-I(u)} \|\phi(x,y)\|_{L^1(\mathbb{R}^{d+3})} \sqrt{\frac{|S^{d+3}|}{d+1}},$$

which is finite as assumed. Thus, $u(x,y) \in L^2(\mathbb{R}^{d+3})$ in the case of the theorem when $a=0$.

Let us conclude the proof by addressing the case b) of the theorem. Let us apply the generalized Fourier transform (15) to both sides of problem (13) and derive

$$\tilde{\tilde{u}}(k,q) = \frac{\tilde{\tilde{\phi}}(k,q)}{\sqrt{k^2+q^2}-a}.$$

We introduce the set

$$B_\delta := \{(k,q) \in \mathbb{R}^{d+3} \mid a-\delta \le \sqrt{k^2+q^2} \le a+\delta\}, \quad 0 < \delta < a. \tag{30}$$

Hence

$$\tilde{\tilde{u}}(k,q) = \frac{\tilde{\tilde{\phi}}(k,q)}{\sqrt{k^2+q^2}-a} \chi_{B_\delta} + \frac{\tilde{\tilde{\phi}}(k,q)}{\sqrt{k^2+q^2}-a} \chi_{B_\delta^c}. \tag{31}$$

Clearly, the second term in the right side of (31) can be estimated from above in the absolute value by $\dfrac{|\tilde{\tilde{\phi}}(k,q)|}{\delta} \in L^2(\mathbb{R}^{d+3})$ due to the one of our assumptions. Evidently, we have the representation

$$\tilde{\tilde{\phi}}(k,q) = \tilde{\tilde{f}}(a,\sigma) + \int_a^{\sqrt{k^2+q^2}} \frac{\partial \tilde{\tilde{\phi}}(s,\sigma)}{\partial s} ds.$$

This allows us to express the first term in the right side of (31) as

$$\frac{\tilde{\tilde{\phi}}(a,\sigma)}{\sqrt{k^2+q^2}-a} \chi_{B_\delta} + \frac{\int_a^{\sqrt{k^2+q^2}} \frac{\partial \tilde{\tilde{\phi}}(s,\sigma)}{\partial s} ds}{\sqrt{k^2+q^2}-a} \chi_{B_\delta}. \tag{32}$$

Apparently, we have the upper bound for the second term in (32) from above in the absolute value by

$$\|(\nabla_k + \nabla_q)\tilde{\tilde{\phi}}(k,q)\|_{L^\infty(\mathbb{R}^{d+3})} \chi_{B_\delta} \in L^2(\mathbb{R}^{d+3}).$$

Note that under our assumptions $(\nabla_k + \nabla_q)\tilde{\tilde{\phi}}(k, q) \in L^\infty(\mathbb{R}^{d+3})$ via Lemma 12 of [19]. It can be easily verified that, the first term in (32) is square integrable if and only if $\tilde{\tilde{\phi}}(a, \sigma)$ vanishes, which is equivalent to orthogonality relation (16).   ∎

We conclude the article with establishing the solvability in the sense of sequences for our problem in the case of a free Laplacian added to a three dimensional Schrödinger operator.

Proof of Theorem 5 Suppose $u(x, y)$ and $u_n(x, y)$, $n \in \mathbb{N}$ are the unique solutions of problems (13) and (17) in $H^1(\mathbb{R}^{d+3})$ with $a \geq 0$ respectively and it is known that $u_n(x, y) \to u(x, y)$ in $L^2(\mathbb{R}^{d+3})$ as $n \to \infty$. Then, it will can be shown that $u_n(x, y) \to u(x, y)$ in $H^1(\mathbb{R}^{d+3})$ as $n \to \infty$ as well. Indeed, from (13) and (17) we easily obtain that

$$L_U(u_n(x, y) - u(x, y)) = a(u_n(x, y) - u(x, y)) + [\phi_n(x, y) - \phi(x, y)].$$

Clearly, this yields

$$\|L_U(u_n(x, y) - u(x, y))\|_{L^2(\mathbb{R}^{d+3})} \leq a\|u_n(x, y) - u(x, y)\|_{L^2(\mathbb{R}^{d+3})} +$$

$$+ \|\phi_n(x, y) - \phi(x, y)\|_{L^2(\mathbb{R}^{d+3})} \to 0, \quad n \to \infty$$

due to our assumptions. Let us express

$$\|L_U(u_n(x, y) - u(x, y))\|_{L^2(\mathbb{R}^{d+3})}^2 = \|\nabla(u_n(x, y) - u(x, y))\|_{L^2(\mathbb{R}^{d+3})}^2 +$$

$$+ \int_{\mathbb{R}^{d+3}} U(y)|u_n(x, y) - u(x, y)|^2 dx dy,$$

where the scalar potential $U(y)$ is bounded via Assumption 1. Hence, in the equality above the left side along with the second term in the right side tend to zero as $n \to \infty$. This implies that $\nabla u_n(x, y) \to \nabla u(x, y)$ in $L^2(\mathbb{R}^{d+3})$ as $n \to \infty$, such that $u_n(x, y) \to u(x, y)$ in $H^1(\mathbb{R}^{d+3})$ as $n \to \infty$ as well.

In the case (a) (13) and (17) admit unique solutions $u(x, y)$, $u_n(x, y)$ belonging to $H^1(\mathbb{R}^{d+3})$ respectively with $n \in \mathbb{N}$ by means of the part (a) of Theorem 4 above. We apply the generalized Fourier transform (15) to both sides of problems (13) and (17). This gives us

$$\tilde{\tilde{u}}(k, q) = \frac{\tilde{\tilde{\phi}}(k, q)}{\sqrt{k^2 + q^2}}, \quad \tilde{\tilde{u}}_n(k, q) = \frac{\tilde{\tilde{\phi}}_n(k, q)}{\sqrt{k^2 + q^2}}, \quad n \in \mathbb{N}.$$

Hence $\tilde{\tilde{u}}_n(k, q) - \tilde{\tilde{u}}(k, q)$ can be expressed as

$$\frac{\tilde{\tilde{\phi}}_n(k, q) - \tilde{\tilde{\phi}}(k, q)}{\sqrt{k^2 + q^2}} \chi_{\{\sqrt{k^2+q^2} \leq 1\}} + \frac{\tilde{\tilde{\phi}}_n(k, q) - \tilde{\tilde{\phi}}(k, q)}{\sqrt{k^2 + q^2}} \chi_{\{\sqrt{k^2+q^2} > 1\}}. \qquad (33)$$

Obviously, the second term in (33) can be trivially estimated from above in the absolute value by $|\tilde{\tilde{\phi}}_n(k, q) - \tilde{\tilde{\phi}}(k, q)|$, such that

$$\left\| \frac{\tilde{\tilde{\phi}}_n(k, q) - \tilde{\tilde{\phi}}(k, q)}{\sqrt{k^2 + q^2}} \chi_{\{\sqrt{k^2+q^2} > 1\}} \right\|_{L^2(\mathbb{R}^{d+3})} \leq \|\phi_n(x, y) - \phi(x, y)\|_{L^2(\mathbb{R}^{d+3})} \to 0$$

as $n \to \infty$ via the one of our assumptions. Let us obtain the upper bound in the the absolute value for the first term in (33) via the Corollary 2.2 of [21] by

$$\frac{1}{(2\pi)^{\frac{d+3}{2}}} \frac{1}{1 - I(U)} \|\phi_n(x, y) - \phi(x, y)\|_{L^1(\mathbb{R}^{d+3})} \frac{\chi_{\{\sqrt{k^2+q^2} \leq 1\}}}{\sqrt{k^2 + q^2}},$$

such that

$$\left\| \frac{\tilde{\tilde{\phi}}_n(k, q) - \tilde{\tilde{f}}(k, q)}{\sqrt{k^2 + q^2}} \chi_{\{\sqrt{k^2+q^2} \leq 1\}} \right\|_{L^2(\mathbb{R}^{d+3})} \leq$$

$$\leq \frac{1}{(2\pi)^{\frac{d+3}{2}}} \frac{1}{1 - I(U)} \|\phi_n(x, y) - \phi(x, y)\|_{L^1(\mathbb{R}^{d+3})} \frac{\sqrt{|S^{d+3}|}}{d + 1} \to 0, \quad n \to \infty$$

via the one of our assumptions. Therefore, $u_n(x, y) \to u(x, y)$ in $L^2(\mathbb{R}^{d+3})$ as $n \to \infty$ when the parameter $a = 0$.

Finally, let us proceed to the proof of the part (b) of the theorem. For each $n \in \mathbb{N}$ problem (17) has a unique solution $u_n(x, y) \in H^1(\mathbb{R}^{d+3})$ via the result of the part (b) of Theorem 4 above. By means of (16) along with Corollary 2.2 of [21], we estimate for $(k, q) \in S_a^{d+3}$

$$\left| \left(\phi(x, y), \frac{e^{ikx}}{(2\pi)^{\frac{d}{2}}} \eta_q(y) \right)_{L^2(\mathbb{R}^{d+3})} \right| = \left| \left(\phi(x, y) - \phi_n(x, y), \frac{e^{ikx}}{(2\pi)^{\frac{d}{2}}} \eta_q(y) \right)_{L^2(\mathbb{R}^{d+3})} \right| \leq$$

$$\leq \frac{1}{(2\pi)^{\frac{d+3}{2}}} \frac{1}{1 - I(U)} \|\phi_n(x, y) - \phi(x, y)\|_{L^1(\mathbb{R}^{d+3})} \to 0, \quad n \to \infty.$$

Note that under our assumptions $\phi_n(x, y) \to \phi(x, y)$ in $L^1(\mathbb{R}^{d+3})$ via the elementary argument on p. 116 of [26]. Thus, we arrive at

$$\left(\phi(x, y), \frac{e^{ikx}}{(2\pi)^{\frac{d}{2}}} \eta_q(y) \right)_{L^2(\mathbb{R}^{d+3})} = 0, \quad (k, q) \in S_a^{d+3}. \qquad (34)$$

Therefore, problem (13) has a unique solution $u(x, y) \in H^1(\mathbb{R}^{d+3})$ via the result of the part (b) of Theorem 4 above. Let us apply the generalized Fourier transform (15) to both sides of equations (13) and (17). This yields

$$\tilde{\tilde{u}}_n(k, q) - \tilde{\tilde{u}}(k, q) = \frac{\tilde{\tilde{\phi}}_n(k, q) - \tilde{\tilde{\phi}}(k, q)}{\sqrt{k^2 + q^2} - a} \chi_{B_\delta} + \frac{\tilde{\tilde{\phi}}_n(k, q) - \tilde{\tilde{\phi}}(k, q)}{\sqrt{k^2 + q^2} - a} \chi_{B_\delta^c} \qquad (35)$$

with the set B_δ defined in (30). Evidently, the second term in the right side of (35) can be estimated from above in the absolute value by $\dfrac{|\tilde{\tilde{\phi}}_n(k, q) - \tilde{\tilde{\phi}}(k, q)|}{\delta}$, such that

$$\left\| \frac{\tilde{\tilde{\phi}}_n(k, q) - \tilde{\tilde{\phi}}(k, q)}{\sqrt{k^2 + q^2} - a} \chi_{B_\delta^c} \right\|_{L^2(\mathbb{R}^{d+3})} \leq \frac{\|\phi_n(x, y) - \phi(x, y)\|_{L^2(\mathbb{R}^{d+3})}}{\delta} \to 0, \quad n \to \infty$$

due to the one of our assumptions. Orthogonality relations (16) and (34) imply that

$$\tilde{\tilde{\phi}}(a, \sigma) = 0, \quad \tilde{\tilde{\phi}}_n(a, \sigma) = 0, \quad n \in \mathbb{N},$$

such that

$$\tilde{\tilde{\phi}}(k, q) = \int_a^{\sqrt{k^2+q^2}} \frac{\partial \tilde{\tilde{\phi}}(s, \sigma)}{\partial s} ds, \quad \tilde{\tilde{\phi}}_n(k, q) = \int_a^{\sqrt{k^2+q^2}} \frac{\partial \tilde{\tilde{f}}_n(s, \sigma)}{\partial s} ds, \quad n \in \mathbb{N}.$$

This allows us to express the first term in the right side of (35) as

$$\frac{\int_a^{\sqrt{k^2+q^2}} \left[\frac{\partial \tilde{\tilde{\phi}}_n(s,\sigma)}{\partial s} - \frac{\partial \tilde{\tilde{\phi}}(s,\sigma)}{\partial s} \right] ds}{\sqrt{k^2 + q^2} - a} \chi_{B_\delta}. \qquad (36)$$

Apparently, (36) can be bounded from above in the absolute value by

$$\|(\nabla_k + \nabla_q)\tilde{\tilde{\zeta}}_n(k, q) - \tilde{\tilde{\phi}}(k, q))\|_{L^\infty(\mathbb{R}^{d+3})} \chi_{B_\delta},$$

which us to estimate the $L^2(\mathbb{R}^{d+3})$ norm of (36) from above by

$$C\|(\nabla_k + \nabla_q)(\tilde{\tilde{\phi}}_n(k, q) - \tilde{\tilde{\phi}}(k, q))\|_{L^\infty(\mathbb{R}^{d+3})} \to 0, \quad n \to \infty$$

by virtue of the part (b) of Lemma 5 of [26] under the given assumptions. This proves that $u_n(x, y) \to u(x, y)$ in $L^2(\mathbb{R}^{d+3})$ as $n \to \infty$. ∎

References

1. Alfimov GL, Medvedeva EV, Pelinovsky DE (2014) Wave systems with an infinite number of localized traveling waves. Phys Rev Lett 112(5):054103
2. Amrouche C, Girault V, Giroire J (1997) Dirichlet and Neumann exterior problems for the n-dimensional Laplace operator: an approach in weighted Sobolev spaces. J Math Pures Appl 76(1):55–81
3. Amrouche C, Bonzom F (2008) Mixed exterior Laplace's problem. J Math Anal Appl 338(1):124–140
4. Bolley P, Pham TL (1993) Propriétés d'indice en théorie höldérienne pour des opérateurs différentiels elliptiques dans R^n. J Math Pures Appl 72(1):105–119
5. Bolley P, Pham TL (2001) Propriété d'indice en théorie Höldérienne pour le problème extérieur de Dirichlet. Comm Partial Differ Equ 26(1–2):315–334
6. Benkirane N (1988) Propriétés d'indice en théorie höldérienne pour des opérateurs elliptiques dans R^n. CRAS 307. Série I(11):577–580
7. Cycon HL, Froese RG, Kirsch W, Simon B (1987) Schrödinger operators with application to quantum mechanics and global geometry. Springer, Berlin, p 319
8. Ducrot A, Marion M, Volpert V (2005) Systemes de réaction-diffusion sans propriété de Fredholm. CRAS 340(9):659–664
9. Ducrot A, Marion M, Volpert V (2008) Reaction-diffusion problems with non-Fredholm operators. Adv Differ Equ 13(11–12):1151–1192
10. Goldberg M, Schlag W (2004) A limiting absorption principle for the three-diensional Schrödinger equation with L^p potentials. Int Math Res Not 75:4049–4071
11. Kato T (1965/1966) Wave operators and similarity for some non-selfadjoint operators. Math Ann 162:258–279
12. Lieb EH, Loss M (1997) Analysis. graduate studies in mathematics. Am Math Soc Providence 14:278
13. Reed M, Simon B (1979) Methods of modern mathematical physics. Academic Press, III. Scattering theory, p 463
14. Rodnianski I, Schlag W (2004) Time decay for solutions of Schrödinger equations with rough and time-dependent potentials. Invent Math 155(3):451–513
15. Volpert V (2011) Elliptic partial differential equations. Volume 1: Fredholm theory of elliptic problems in unbounded domains. Birkhauser, p 639
16. Volpert V, Kazmierczak B, Massot M, Peradzynski Z (2002) Solvability conditions for elliptic problems with non-Fredholm operators. Appl Math 29(2):219–238
17. Volpert V, Vougalter V (2011) On the solvability conditions for a linearized Cahn-Hilliard equation. Rend Istit Mat Univ Trieste 43:1–9
18. Vougalter V (2013) Sharp semiclassical bounds for the moments of eigenvalues for some Schrödinger type operators with unbounded potentials. Math Model Nat Phenom 8(1):237–245
19. Vougalter V, Volpert V (2010) On the solvability conditions for some non Fredholm operators. Int J Pure Appl Math 60(2):169–191
20. Vougalter V, Volpert V (2010) Solvability relations for some non Fredholm operators. Int Electron J Pure Appl Math 2(1):75–83
21. Vougalter V, Volpert V (2011) Solvability conditions for some non-Fredholm operators. Proc Edinb Math Soc 54(1):249–271
22. Vougalter V, Volpert V (2011) On the existence of stationary solutions for some non-Fredholm integro-differential equations. Doc Math 16:561–580
23. Vougalter V, Volpert V (2012) On the solvability conditions for the diffusion equation with convection terms. Commun Pure Appl Anal 11(1):365–373
24. Vougalter V, Volpert V (2012) Solvability conditions for a linearized Cahn-Hilliard equation of sixth order. Math Model Nat Phenom 7(2):146–154

25. Vougalter V, Volpert V (2012) Solvability conditions for some linear and nonlinear non-Fredholm elliptic problems. Anal Math Phys 2(4):473–496
26. Vougalter V, Volpert V (2014) On the solvability in the sense of sequences for some non-Fredholm operators. Dyn Partial Differ Equ 11(2):109–124
27. Vougalter V, Volpert V (2017) Existence of stationary solutions for some integro-differential equations with superdiffusion. Rend Semin Mat Univ Padova 137:185–201

Printed in the United States
by Baker & Taylor Publisher Services